+ KJARTAN POSKITT %

EVERYDAY MATHS FOR GROWN-UPS

Michael O'Mara Books Limited

First published in Great Britain in 2010 by
Michael O'Mara Books Limited
9 Lion Yard
Tremadoc Road
London SW4 7NQ

A CIP catalogue record for this book is available from the British Library.

Papers used by Michael O'Mara Books Limited are natural, recyclable products made from wood grown in sustainable forests. The manufacturing processes conform to the environmental regulations of the country of origin.

ISBN: 978-1-84317-384-7

1 2 3 4 5 6 7 8 9 10

www.mombooks.com

Cover design by Allan Somerville
Designed and typeset by RefineCatch
Illustrations by Andrew Pinder

Printed and bound in England by Clays Ltd, St Ives plc

CONTENTS

Making a Rough Guess

Fractions

Ratios

Decimals

Powers and Roots

Averages

Algebra

For Marilyn Malin, who has been keeping me organized for over twenty years, and who has never used a calculator and is never wrong.

HOW THIS BOOK CAME TO BE WRITTEN

Not so long ago a mate of mine approached me, looking desperate. Blakey is about forty, and as intelligent as anybody else I know, but he was having trouble getting onto a management course because he kept failing the numeracy exam. In his own words: 'I can just about add and take away, but when it gets to multiplying, I lose it: even with a calculator I don't know if I've done it right.' So I lent Blakey a copy of *The Awesome Arithmeticks,* which I'd written for eight-year-olds, and within a few weeks he passed.

If you are one of the many people, like Blakey, who say you can't do sums, it's probably because you missed something vital early on, and after that nothing made sense. That's why I decided to write a book that starts with how to add, then works up to the fancy bits so you can follow it all through and see how everything slots together. If the early parts of the book seem too easy then skip them, but if you get stuck later you can always look back and see where you lost the plot.

Don't worry: this is *not* a textbook! Yes, of course there are lots of numbers and diagrams and a few nerdy things like π and

x^2 and so on, but there are no tests or exams and nobody is going to shout at you if you fall asleep reading it. This book is here to give you some friendly advice with everyday sums, such as working out how much paint you need for a room, or planning how long a journey should take. It also guides you through some of the weirder stuff such as algebra and percentages, so that you can't be intimidated by twelve-year-olds discussing their maths homework. Along the way we'll look at some fun things like curved space and poker hands, and there are even a few tricks to show off to your friends!

Here's a trick to get you started (use a calculator if you like):

- ✔ Pick any three-digit number. The digits should be different.
- ✔ Turn it round.
- ✔ Subtract one number from the other.

724	or	564
- 427		- 465
= 297		= 099

The answer will always have 9 in the middle (or it will be 99) and the first digit and last digit will always add up to 9!

If you've got an impressionable friend called Malcolm, you can amaze him with this trick. Get Malcolm to write down a three-digit number without telling you what it is. (The digits must all be different.) Then tell him to turn the number round, and subtract one number from the other. Ask him to tell you the first digit in the answer: even though you don't know what numbers he started with, you can tell him what the rest of his answer is!

If Malcolm says the first digit is 9, the answer is 99. Otherwise you can quickly work it out: if Malcolm says the first digit is 5, the answer is 594. Remember, there will always be 9 in the middle and the first digit and last digit will add up to 9!

ADDING

Adding is one of the first things you ever learn to do at school, but don't take it for granted! If it seems simple, that's because we use the ingenious Hindu-Arabic number system, which can handle numbers of any size, but only uses ten digits: 0, 1, 2, 3, 4, 5, 6, 7, 8 and 9. Here's a reminder of how it works.

Using the Place System

You have had three absolutely amazing days at the car-boot sale. The different days' takings were £173, £585 and £234. The only irritating thing is that you accidentally sold your calculator. So how much did you take altogether?

The digits use a 'place' system so if you see the number 173, you know the 3 stands for three units, the 7 for seven tens and the 1 for one hundred. When you get a sum such as 173 + 585 + 234, all you need to do is write the numbers out so that hundreds, tens and units are in columns.

		hundreds	tens	units
Start adding the UNITS		1	7	3
3 + 5 + 4 = 12	+	5	8	5
Put the 2 in the answer and mark the 1 in the tens column	+	2	3	4
			1	
				2

		hundreds	tens	units
Next add the TENS including the extra ten we got from adding the units		1	7	3
7 + 8 + 3 + 1 = 19	+	5	8	5
Finally add the HUNDREDS	+	2	3	4
1 + 5 + 2 + 1 = 9		1	1	
		9	9	2

Hindu-Arabic Versus Roman

We use the Hindu-Arabic number system, which evolved in India around 2,400 years ago. Around 1,100 years ago, it was adopted by Arab mathematicians and astronomers, and was finally introduced to Europe by Leonardo Fibonacci of Pisa about 800 years ago (about the same time as the famous Leaning Tower of Pisa was being built).

It's hard to appreciate how clever this number system is until you imagine doing the same sum using ancient Roman numerals. The Romans used letters as follows:

M = 1000	X = 10
D = 500	V = 5
C = 100	I = 1
L = 50	

Most of the time, numbers were put together by using mixtures of the letters above, starting with the highest value and working down to the smallest. For example, CLXXIII = 100 + 50 + 10 + 10 + 1 + 1 + 1 = 173. However it got tedious writing numbers such as 9, which would have been VIIII, therefore if a smaller value letter was written *before* a bigger value, it was subtracted, so 9 could instead be written as IX.

Roman numerals are still used by people who want to make something look a bit stylish or classy. Old-style clock faces use I to XII to mark the numbers 1 to 12 and a lot of films and TV programmes use Roman numerals to show the copyright year at the very end of the closing titles – so MMX represents the year 2010. Grand buildings and statues often have a foundation stone carved with the date using Roman numerals: the Statue of Liberty in New York Harbor holds a tablet inscribed with the date of the US Declaration of Independence as JULY IV MDCCLXXVI (July 4th 1776).

from Nero to Zero

The ancient Romans didn't have a symbol for 0. It was only when we all started using the place system for writing numbers such as 10 and 100 that 0 became important.

The one place you won't see Roman numerals these days is in sums. Imagine trying to work out the income from your car-boot sale in Roman times . . .

With the Hindu-Arabic system, so long as you make sure that the units, tens and hundreds are lined up in columns, the numbers do the hard work for you. What's more, if you get into the habit of jotting sums down, you'll develop an instinct for the right answer that you'll never get if you always rely on a calculator!

Sometimes the sums are already written out for you, so here's a way to put your instincts into practice . . .

How to Check a Till Receipt Quickly

Have you ever come out of a shop clutching a long till receipt and thinking you've been overcharged? As you struggle along with everything bursting out of the bags, you really don't want to stop and spend ten minutes trying to add it all up, but luckily that's not necessary. There's a way to get a reasonably accurate total very quickly.

Here's a receipt with the total torn off. There are just two things to do.

Skinty's Stores	
Cheese	2·79
Washing Powder	4·35
Newspaper	·40
Dog Grooming Set	6·20
Cornflakes	2·30
Squash	1·49
Eggs	1·20
Bottle of Plonk	5·79
Paint	3·15
Sausages	2·69
Garlic	1·30
Gas Mask	7·49
Plastic Flowers	3·00
Batteries	3·89
Something that needs batteries	4·80
DVD "The Best of Newsnight"	11·49
Spoon	·45
Bath Salt gift set	2·30
Bananas	1·56

❶ Add up the pounds (and ignore the pennies). Here we get a total of £58.

❷ Fold the list of items in half, then add on £1 for every item showing.

There are 10 items here so we add £10 to £58 and get a total of £68. This should be quite close to the real total.

Skinty's Stores	
Cheese	2·79
Washing Powder	4·35
Newspaper	·40
Dog Grooming Set	6·20
Cornflakes	2·30
Squash	1·49
Eggs	1·20
Bottle of Plonk	5·79
Paint	3·15
Sausages	2·69

Let's have a look . . .

TOTAL	66·64

We estimated £68 so that's not bad!

Two Things to Watch Out For

If the shop has 'multipack' offers or other discounts, there might be some negative numbers in the list. Try to ignore these and then subtract at the end. Also, if you're filling a selection of bags, some shops give a 'bag total' as each bag is filled, so you have to ignore these too.

Why Does it Work?

The numbers in the pence column can range between 0p and 99p. Some items will only have a low number of pennies (e.g. 25p) whereas others might have a high number (e.g. 80p). The numbers average out at around 50p per item so to get a rough total of the pennies, we could just count up the items and add 50p for each one. However, it's much easier to cut the number of items in half (that's why we folded the receipt) and add £1 per item.

More shopping tips! There's a whole section on percentages, savings and discounts on pages 98–99.

SUBTRACTING

Although you can add several numbers together at once, you should never try to subtract more than one number at a time. First of all we'll look at the old way of subtracting, and then we'll see the groovy new way that some kids are taught to do it.

The Old Way

The key to subtracting is knowing that if you have a number such as 73, it's the same as 70 + 3.

If we need to work out 73 – 2 it's easy. We just need to subtract the units at the end to get 3 – 2 = 1. We haven't needed to touch the 70 part, so it appears unchanged in the answer. (It helps to use squared paper, to keep track of which are tens and which units, and so on.)

```
  7 3
-   2
= 7 1
```

The fun starts when we have 73 – 9. This is the same as 70 + 3 – 9, but we can't work out the 3 – 9 bit quite so easily.

```
  7 3
-   9
=
```

What we have to do is break the 73 up into 60 + 13. This involves changing the 7 to a 6, and putting a little 1 in front of the 3. This is why I've used squared paper; it's to show you that the number on the top is 60 + 13, *not* 613.

```
  6
  7 13
-    9
=
```

We then work out 13 – 9 = 4, and that's the units finished with! Out of our 70, we've only got 60 left, so our final answer is 60 + 4 = 64.

```
  6
  7 13
-    9
= 6  4
```

Now we've got the basic idea, let's not mess about. Suppose you're planning to make a model battleship from 6,305 matchsticks. So far you've only collected 1,847 so how many more do you need?

Here's the sum, and the trick is to start at the units end and work along. The first problem we hit is working out 5 – 7. We need to pinch a ten from somewhere, but 6,305 has a zero in the tens column, so we move along the top a bit further to include the 3. Now we can get the extra ten we need by working out 30 – 1 = 29.

	6	3	0	5
-	1	8	4	7
=				

		2	9	
	6	~~3~~	~~0~~	15
-	1	8	4	7
=				8

Here you'll see that we've replaced the 30 with 29, and also put a 1 in front of the 5. Now we can work out 15 – 7 = 8.

As we've finished sorting out the units, we can block them off and concentrate on the rest of the sum. We're left with 629 – 184. As 9 – 4 = 5, we can put that straight in without any fuss, and that's the tens column sorted out.

		2	9
	6	~~3~~	~~0~~
-	1	8	4
=			5

All that's left to deal with is 62 – 18.

As we can't do the 2 – 8, we take 1 off the 6, leaving 5, and shove the 1 in front of the 2. This gives us 12 – 8 = 4, and finally 5 – 1 = 4 at the front.

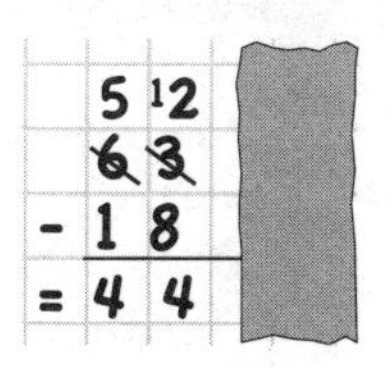

Here's what the completed sum looks like:

	5	12	9	
	~~6~~	~~3~~	~~0~~	15
-	1	8	4	7
=	4	4	5	8

So now you know that you need 4,458 more matchsticks to complete your battleship. (Either that or you need another hobby.)

The New Way

These days, kids are taught to subtract by starting with the smaller number and adding on until they reach the bigger number. It's the same thing Janet does when she gives you your change in Lardy's Pie Shop. If you give her £5 for a pie costing £2·23, the change should be £5 – £2·23 = £2·77. To make sure she's worked it out right, Janet gives you a running commentary: she begins by saying the price of the pie, then adds on the value of each coin as she hands it over, starting with the smallest first, until she reaches £5.

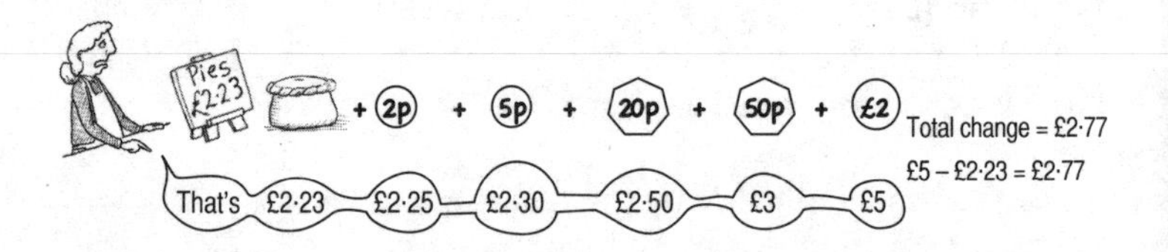

You can use a similar method to do subtractions. Let's look at our matchsticks again. We need to know the answer to 6,305 – 1,847. Let's start by adding bits on to 1,847 and keeping track of them as we go:

1,847 + 3 = 1,850 ➜ so far we've added 3
1,850 + 50 = 1,900 ➜ so far we've added 53
1,900 + 100 = 2,000 ➜ so far we've added 153
2,000 + 4,000 = 6,000 ➜ so far we've added 4,153
6,000 + 300 = 6,300 ➜ so far we've added 4,453
6,300 + 5 = 6,305 ➜ we've got there, and in total we added 4,458

So that's the answer: 6,305 – 1,847 = 4,458. It looks like there's a lot of numbers splashing about, but with a bit of practice you should get the hang of it. Neat, isn't it?

Negative Numbers

A negative number always has a minus sign in front of it. We don't bother putting the '+' sign in front of positive numbers unless it's in a sum such as 3 + 6 – 4 = 5. Here the 3, 6 and 5 are all positive and the 4 is negative.

All numbers are either positive (+) or negative (-).

Sometimes a sum can give a negative answer, especially where money is involved:

Any number that is owed is being taken away, so it is negative.

It can be confusing taking a bigger number away from a smaller number. It helps to imagine a number line with zero in the middle. Positive numbers go off one way and negative numbers go the opposite way.

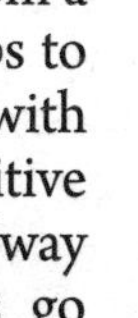

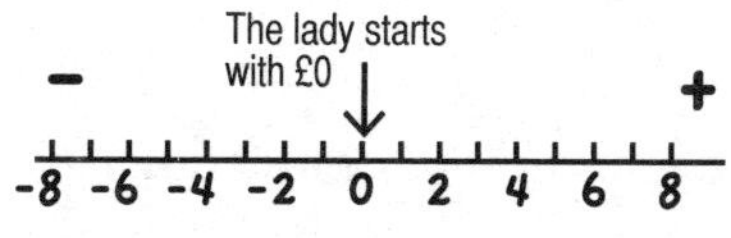

When the lady finds £5, this moves her five steps along in the positive direction.

But when the kid asks for £7, this pulls her back across the zero and she finishes up on the negative end of the line. She has handed her £5 over and she still owes another £2.

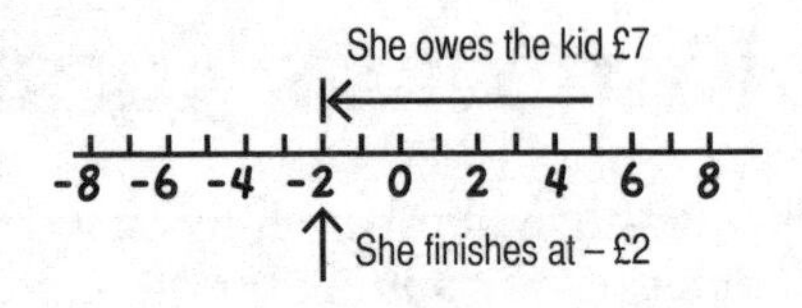

With bigger numbers it isn't always so obvious how much more you might still owe. Suppose you're playing Monopoly and you have £623. You land on Piccadilly, which has four houses on it. You owe a rent of £1,025. You hand over all your cash, but of course you don't have enough to pay the full amount. How much more do you still have to pay? The sum is £623 – £1,025.

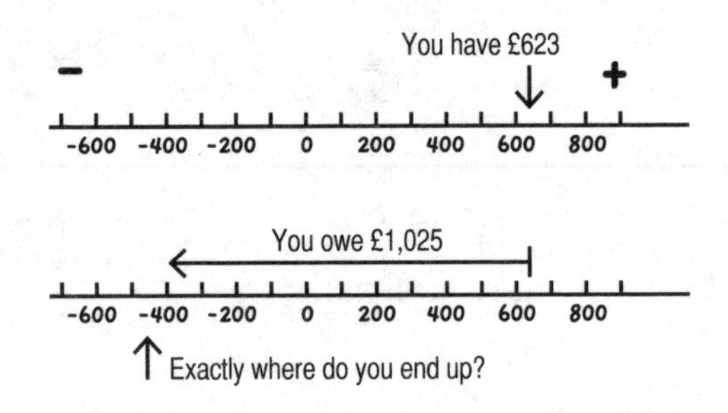

It helps to split subtraction into two separate jobs:

❶ If your negative number is bigger, then your answer is going to be negative. So make sure there's a minus sign at the front when you've finished.

❷ Find the *difference* of the two numbers. This means you have to subtract the smaller number from the bigger number. The difference between 623 and 1,025 is 1,025 – 623 = 402.

Don't forget the minus sign! The answer is – £402, so that's how much you still have to pay. Either that, or you just chuck the Monopoly board across the room and watch all those little bits of plastic and paper go flying everywhere. You won't be loved, but it'll feel good all the same.

MULTIPLYING

'Three times seven is twenty-one, four times seven is twenty-eight . . . ' Let's be honest here: learning the times tables off by heart is one of the most tedious things you ever have to do, but unfortunately they are far too useful to ignore. However the times tables become far friendlier if you know some of the tricks, short cuts and other secrets of how the numbers all lock together.

The Hidden Secrets of the Times Tables

This grid shows all the results for 1 × 1 up to 10 × 10. There are 100 answers on here, so first of all let's get rid of some:

If you multiply by 10, all you do is put a zero on the end of your number. It's very easy, and when we get on to big multiplications, we don't even need it. So we'll get rid of the 10 row and the 10 column.

If you turn a multiplying sum round, you get the same answer, for example both 3 × 7 and 7 × 3 come to 21. So we can get rid of a lot more answers.

	1	2	3	4	5	6	7	8	9	10
1	1	2	3	4	5	6	7	8	9	10
2	2	4	6	8	10	12	14	16	18	20
3	3	6	9	12	15	18	21	24	27	30
4	4	8	12	16	20	24	28	32	36	40
5	5	10	15	20	25	30	35	40	45	50
6	6	12	18	24	30	36	42	48	54	60
7	7	14	21	28	35	42	49	56	63	70
8	8	16	24	32	40	48	56	64	72	80
9	9	18	27	36	45	54	63	72	81	90
10	10	20	30	40	50	60	70	80	90	100

By now we've got rid of well over half the grid. Let's see what's left:

The grey numbers are known as *square* numbers or just squares. These are the answers you get when you multiply any number by itself. For instance, a chess board has 8 squares along each side, so the total number of squares on the board is *eight squared*. We write this as 8^2 which is the same as $8 \times 8 = 64$.

	1	2	3	4	5	6	7	8	9
1	1								
2	2	4							
3	3	6	9						
4	4	8	12	16					
5	5	10	15	20	25				
6	6	12	18	24	30	36			
7	7	14	21	28	35	42	49		
8	8	16	24	32	40	48	56	64	
9	9	18	27	36	45	54	63	72	81

64 little squares can form one big square because 64 is a square number

You can't form a perfect square shape from 63 or 65 squares because these are not square numbers

If you have a phobia about times tables, you can fill in the tables grid another way. First you can fill in all the square numbers just by adding the odd numbers 1, 3, 5, 7 and so on. You start with 1 then you add 3 to get 4. Then add 5 to get 9, then add 7 to get 16 . . . and all the square numbers start turning up.

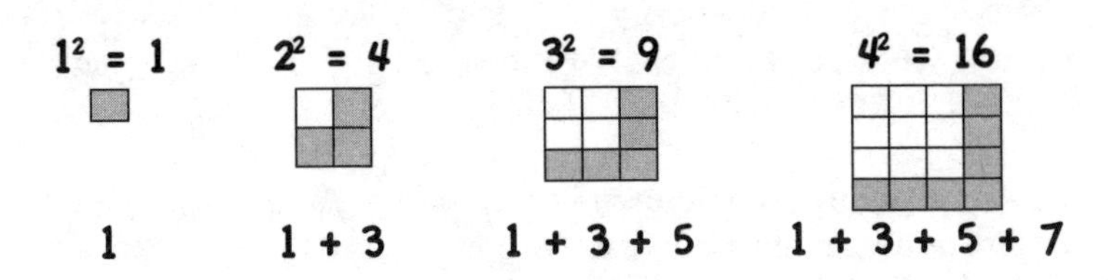

If you pick any of the square numbers on the grid and then subtract the odd numbers starting with 1, you get the numbers on the diagonal going the other way.

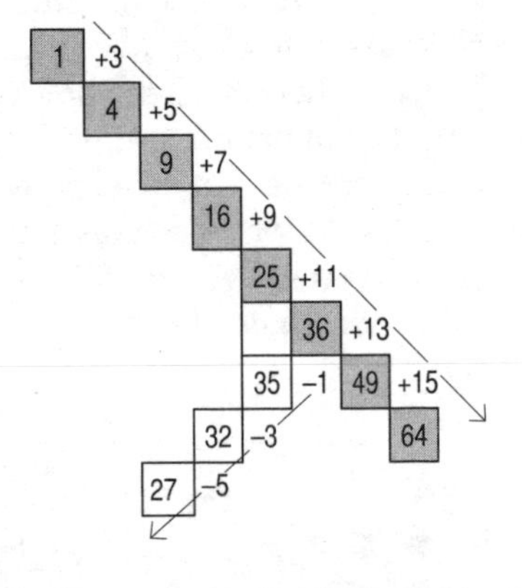

So if we start with 36, subtract 1 to get 35, then subtract 3 to get 32, then subtract 5 to get 27.

(If you compare this diagram to the tables grid, you'll see how the numbers all fit in.)

You can fill in the rest of the grid in exactly the same way, just using the even numbers (2, 4, 6, 8 . . .). Look at the diagonal running underneath the squares which goes 2, 6, 12, 20: you can fill this in by starting with 2, then adding 4, then 6, then 8 and so on. And then if you pick any of *these* answers (e.g. 20), you can subtract 2, then 4, then 6 to get the diagonal going the *other* way (e.g. 20 – 2 = 18 then 18 – 4 = 14 and 14 – 6 = 8).

Using these sequences of odd and even numbers, you can go on to create a set of times tables as big as you like, and you don't need to do any multiplying!

The Three-Number Trick

Think of any three consecutive numbers: the first and last numbers always multiply to make one less than the middle number squared.

If we pick 6, 7, 8 and check the times tables, we find that $6 \times 8 = 48$ and 7×7 (or 7^2) $= 49$.

This works for *any* three numbers in a row. If you just happened to know that $148^2 = 21{,}904$, you also know that $147 \times 149 = 21{,}903$.

(Why does this happen? It's one of those little mysteries we can solve when we get to algebra! See page 85.)

Prime Numbers

A prime number will only divide by itself and 1. For instance, 10 is not prime (it divides by 1, 2, 5 and 10) and 12 is not prime (it divides by 1, 2, 3, 4, 6, 12) but 11 is prime (it only divides by itself and 1). If you want to try and pack things neatly in crates without wasting any space, prime numbers are a nuisance because you can't split them neatly.

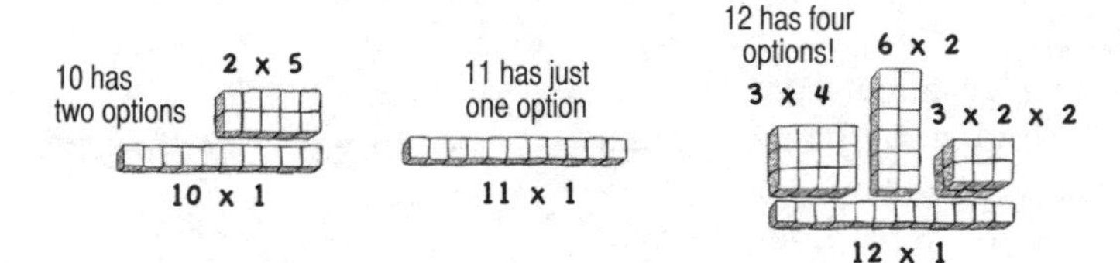

The smallest prime number is 2. It's also the only even prime number because all the other even numbers will divide by 2. The next prime numbers are 3, 5, 7, 11, 13, 17, 19, 23 . . . the list goes on for ever.

Here are all the numbers up to 100 with the primes highlighted. It's easy to see where primes definitely *won't* appear – because after the top row the primes cannot end in 2, 4, 6, 8 or 0 (as they'd divide by 2) and they can't end in 5 (they'd divide by 5). What nobody has ever been able to do is predict where they definitely *will* appear. At one point they got terribly excited because 31 is prime, and so is 331 and 3,331 and 33,331 and 333,331 and so on. It looked like any row of 3s followed by a 1 would be prime until somebody ruined it by working out that 19,607,843 × 17 = 333,333,331. Incidentally if YOU manage to find a pattern in the primes, your name will be remembered long after every celebrity currently infesting the planet has long been forgotten.

Experts can't decide if 1 is prime or not. Normal people don't care . . .

1	2	3	4	5	6	7	8	9	10
11	12	13	14	15	16	17	18	19	20
21	22	23	24	25	26	27	28	29	30
31	32	33	34	35	36	37	38	39	40
41	42	43	44	45	46	47	48	49	50
51	52	53	54	55	56	57	58	59	60
61	62	63	64	65	66	67	68	69	70
71	72	73	74	75	76	77	78	79	80
81	82	83	84	85	86	87	88	89	90
91	92	93	94	95	96	97	98	99	100

Finger Tables

One of the trickier times tables is the 9 times, but almost any kid these days will be able to show you a very neat way of getting it right.

Hold your hands up and imagine your fingers are numbered 1–10 from left to right. Fold down the finger you want to multiply 9 by. See how many fingers are to the left and right of the folded finger. This gives the answer as the diagram shows.

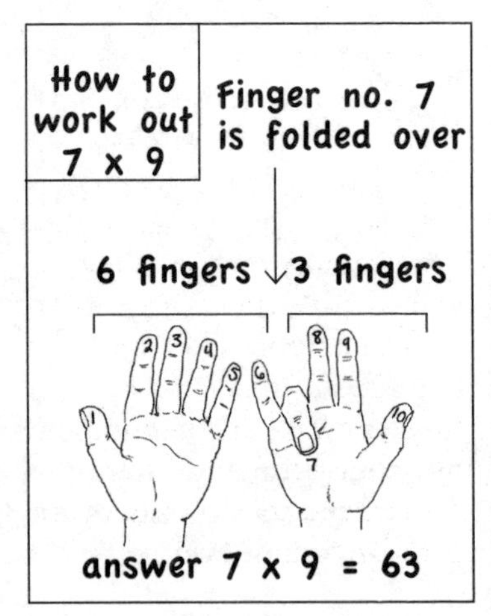

Here's something even stranger . . .

If you know the tables up to 5 × 5, you can do any sum from 6 × 6 up to 10 × 10 on your fingers. First imagine your fingers are numbered 6, 7, 8, 9, and 10 on each hand.

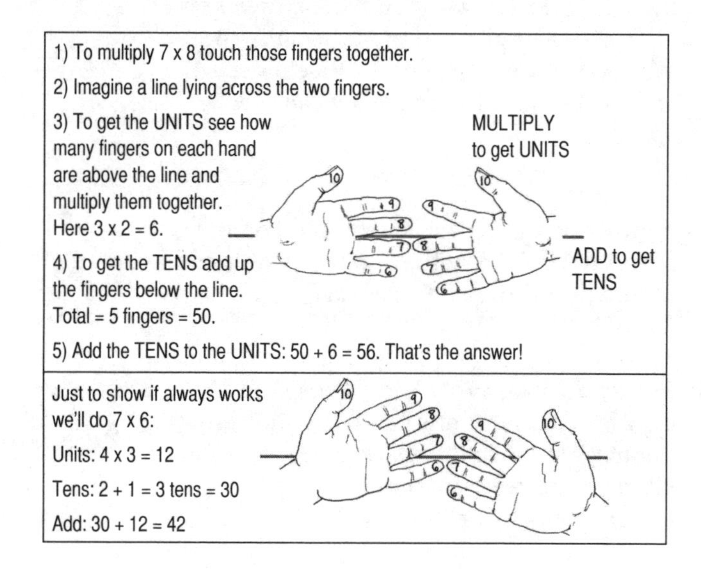

Big Multiplication

You've driven 693 miles to go camping in the middle of nowhere, but when you get back you realize that your front door key must have fallen out of your pocket when you were packing away the tent. By the time you've gone back and found it, you've done the route four times. How far have you travelled?

To be honest, you're probably not in the mood to do sums, but if by any mad chance you do want to work it out, obviously we've now moved beyond the times tables. The trick is to multiply one little bit at a time, and the good news is that you'll never need to multiply by anything bigger than 9. Here's the complete guide to doing 693 × 4.

❶ Set the sum out like this:

```
  693
×   4
=
```

❷ You're going to multiply the 3, then the 9, and then the 6 by 4, making sure the answers end up in the right place. Start at the units end. Work out 3 × 4 = 12. Put the 2 below the 4 and write a little 1 over the next answer space.

```
  693
×   4
=  ¹2
```

Here's how the 12 fits in.

❸ Now work out 9 × 4 = 36, then add on the little 1 to get 37. Put the 7 in the answer and write a little 3 over the next space.

```
  693
×   4
= ³7¹2
```

Here's the 37.

❹ Finally do 6 × 4 = 24. Add on the little 3 to get 27. There's nothing else to multiply, so put in 27 to finish the answer! Feel smug. (It might make up for how daft you felt for leaving your key behind.)

```
   693
×    4
   3 1
= 2772
```

Now we'll move on to multiplying by bigger numbers. If you've got to work out 517 × 38, the old-fashioned way to do it is to multiply 517 × 30 and then 517 × 8, and then add the answers together. It's ugly, but it works.

❶ Write the sum out like this, with some extra lines underneath. First you multiply 517 × 30. Start by putting a 0 under the 8. This makes sure that the rest of the answer goes in the right place.

```
  517
×  38
    0
_____
_____
_____
```

❷ Now you multiply 517 × 3. Start with 7 × 3 = 21. The 1 goes in the same column as the 3, and then write a little 2 in the next column. Don't forget it's 1 × 3 = 3 next! (It's easy to miss a digit out if you're not thinking.) Add the 2 to the 3 to get 5, then put 5 as the answer. Finally 5 × 3 = 15, which goes at the front.

```
  517
x  38
   2
15510
```

❸ Now work out 517 × 8 and put the answer on the lower line. When you first do 7 × 8 = 56, the 6 goes in the same column as the 8.

```
  517
x  38
   2
15510
  1 5
 4136
```

❹ Once you've worked out 517 × 30 and 517 × 8, you add the two results. You get 15,510 + 4,136 = 19,646 and that's the final answer!

```
  517
x  38
   2
15510
  1 5
 4136
19646
```

The Foolproof Way to Multiply

This way takes a bit more planning than the old method but it ensures that all the correct numbers are multiplied together, and all the answers end up in the right columns.

To work out 517 × 38 you start by drawing a grid of boxes with diagonal lines running across them, like this. Write the numbers to be multiplied along the top and down the side.

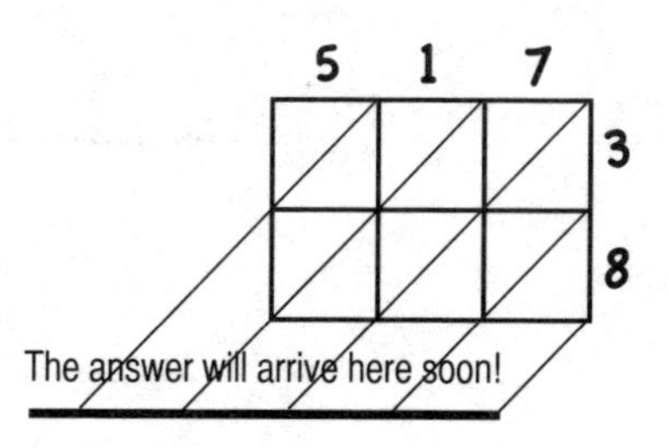

Fill in each little box by multiplying the digit at the top and the digit at the side. For instance, to fill in the top left corner box we worked out 5 × 3 = 15. The 15 goes in the box with the 1 above the diagonal and the 5 underneath.

If any of the sums give a single digit answer (such as 1 × 3 = 3) you fill it in as 03, with a 0 above the diagonal and a 3 underneath.

When you've filled in all the boxes, you just add the numbers up along the diagonals. (Notice that 8 + 5 + 1 = 14, so 4 goes at the bottom and there's a little 1 to be added in the next space along.)

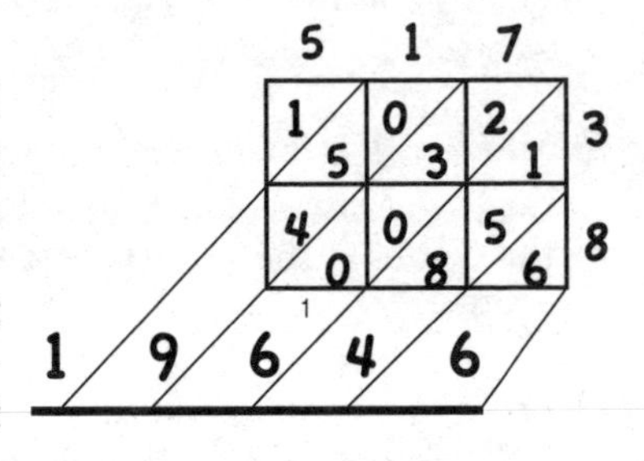

Old-fashioned people might prefer the simpler layout, but even they will probably admit that when it comes to ugly decimals* such as 64·29 × 27·3 things can easily end up in the wrong place. With these grids, it's easy:

To find out where the decimal point goes in the final answer, see where the lines from the first two points intersect, then follow the diagonal all the way down.

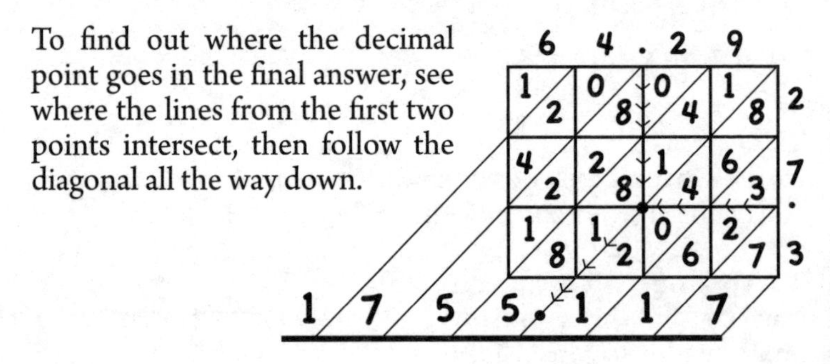

* If the word *decimals* sent you into a panic, don't worry. It's all explained later on.

If only one number has a decimal point, put that number down the right hand side then follow the diagonal from the decimal point straight down from the edge of the grid.

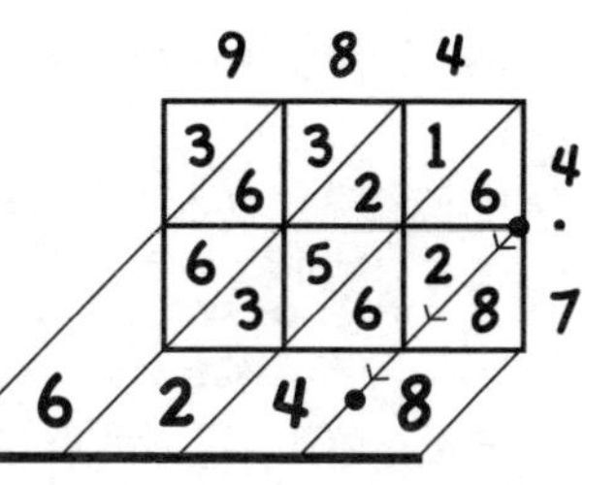

Multiplying Hundreds and Thousands

What's 3,000 × 900? Simple: just multiply the digits at the front (3 × 9 = 27) and then add up the total number of zeros on the end of the numbers in the sum and plonk them on the end. Here we've got five zeros to add, giving us 2,700,000.

You just need to be a bit careful with something like this: 7,500 × 80. First you multiply 75 × 8 = 600. You then need to add on three *extra* zeros because that's how many there were in the sum to start with. The answer is 600,000.

If you have 1,030 × 50, you first work out 103 × 5 = 515. You then just add on two zeros to get 51,500. You don't count the zero in the middle of 103, because you've already allowed for that one when you worked out 103 × 5.

Multiplying Negative Numbers

If only *one* of your multiplying numbers is negative, the answer is negative. If you are multiplying *two* negative numbers, the answer is positive.

$3 \times 2 = 6$ $\quad$ $3 \times -2 = -6$ $\quad$ $-3 \times 2 = -6$ $\quad$ $-3 \times -2 = +6$

Why Does Negative X Negative = Positive?

The easiest way to explain this is with a nice sensible example... Imagine you're running home from the bus stop in a thick fog. Your speed is +3mph. This is a positive speed so after 1 hour you're 3 miles nearer home.

Now imagine you're facing the wrong way and running at 3mph. This is a negative speed of −3mph and after 1 hour you're 3 miles further from home.

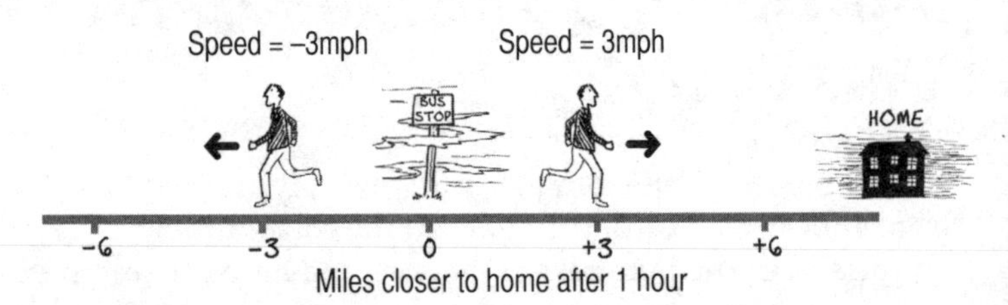

We'll start at the bus stop again, but this time you decide to run twice as fast. If you're facing the right way, you'll be running at $3 \times 2 = 6$mph. If you're facing the wrong way you are running at $-3 \times 2 = -6$mph.

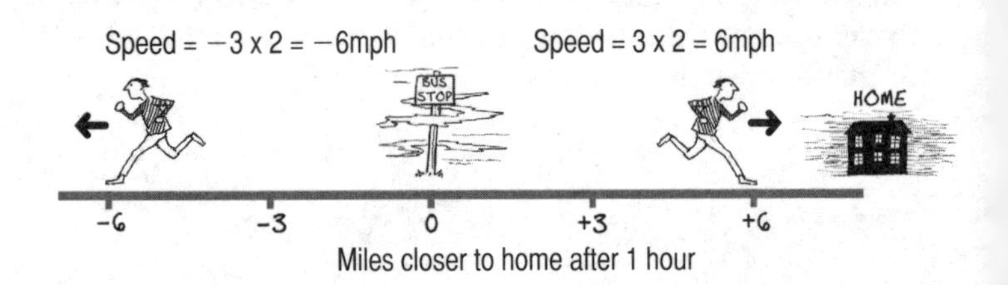

Let's start at the bus stop one last time but, remember, it's still thick fog and this time you think you might be facing the wrong way. Therefore as well as doubling your speed, you're going to run *backwards*. This is like multiplying your speed by −2. If you were facing towards home, you'd be running away from it twice as fast and so your speed would be

$3 \times -2 = -6$mph. But if you were facing away from home and running backwards, your speed would be $-3 \times -2 = +6$mph. You would actually be running home twice as fast!

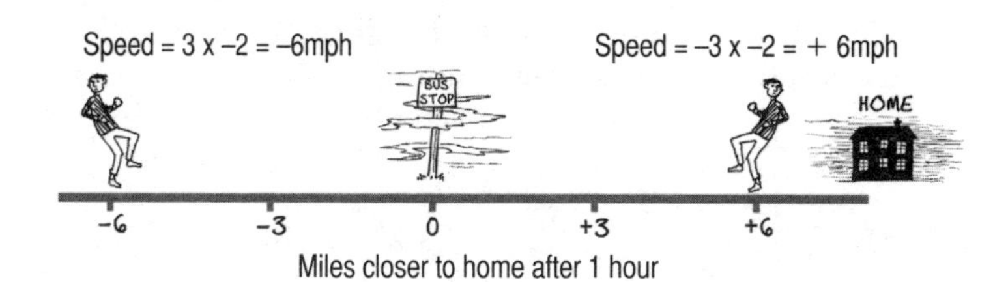

(Please note: we have just proved that facing the wrong way and running backwards in thick fog is as good as facing the right way and running forwards. However, this is only a mathematical example. The author and publishers can take no responsibility for any injuries caused by... oh you get the general idea.)

It isn't just sums in which negatives cancel each other out. Sometimes newspaper reports have so many negatives that you don't know what they are saying. Try this:

Mrs Beaumont denied that she had refused to object to the appeal against overturning the ban on footballers ripping their shirts off.

So does Mrs Beaumont approve of topless footballers or not? I'll give you a clue: Mrs Beaumont keeps her season ticket in the same bag as her binoculars.

Three Calculator Tricks

❶ Get a calculator and tap in 12345679 (missing out 8). Now multiply it by any answer in the 9 times table (e.g. 9 or 18 or 27...).

❷ Pick any number between 100 and 999 and tap it into a calculator. Now multiply it by pressing these buttons: × 7 × 11 × 13 =

❸ Pick any number between 10 and 99 and tap it into a calculator. Now press these buttons: × 3 × 7 × 13 × 37 =

And if You Only Know the Two Times Table . . .

Amazing, but true! You can multiply *any* two numbers just by multiplying and dividing by two. This is called 'Russian Peasant Multiplying', although the ancient Egyptians also used it, some computer systems still use it and, for all we know, alien life forms simply can't live without it. Here's how to multiply 326 by 28 only using the two times table:

1) Write the numbers to be multiplied (326 and 28) at the top of the page and draw a vertical line between them.

2) Divide the first number (326) by two, ignoring any remainders, and write the answer below. Keep dividing by two and writing the answer below until you finally get to 1.

326	~~28~~
163	56
81	112
40	~~224~~
20	~~448~~
10	~~896~~
5	1792
2	~~3584~~
1	7168
	= 9128

3) Keep multiplying the second number (28) by two, putting your answers alongside the numbers in the first column, until you reach the 1.

4) Cross out the numbers in the right-hand column that are opposite an even number in the left-hand column.

5) Add up the remaining numbers in the second column.

That's it!

DIVIDING

If it wasn't for dividing, maths would be so much simpler. If we just pick two little numbers and try to + – × and then ÷ we can see why. Let's take 9 and 7:

$9 + 7 = 16$ . . . *easy!* $9 - 7 = 2$ . . . *easy!*
$9 \times 7 = 63$. . . *no big deal.*

But when we try $9 \div 7$, there are at least three ways of giving the answer:

$9 \div 7 = 1$ remainder 2 (or 1 with 2 left over)
$9 \div 7 = 1\frac{2}{7}$ (or in some cases you can just leave it as $\frac{9}{7}$)
$9 \div 7 = 1{\cdot}2857142857$ . . . *ugh!*

Sharing Out

We won't worry about fractions or decimals yet; we'll start very humbly with whole numbers and then, when we're sure what's happening, we'll gradually turn the heat up.

You've got 8 kids at a birthday party and 24 buns. How many buns does each kid get? The sum is: how many *times* does *eight* go into twenty-four? The words *eight* and *times* are telling you to look at the 8 times table and find that $8 \times 3 = 24$. This tells us that $24 \div 8 = 3$ so each kid gets three buns. Some people are already confused at this stage, so we'll lay it out with dots:

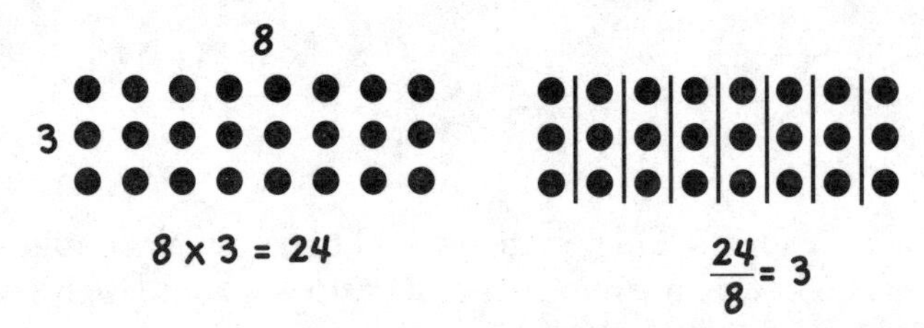

If you split 24 into 8 bits, there will be 3 in each bit.

Dividing Bigger Numbers

That wasn't too tough because 24 appears in the 8 times table. Life gets trickier when your 8 kids have 53 balloons to share out. How many times does 8 go into 53?

Unfortunately, 53 doesn't appear in the eight times table so we need to find the answer that's closest to 53 without going over. The sum we need is $8 \times 6 = 48$, which tells us that each kid can have 6 balloons and that will get rid of 48 of them. If we work out $53 - 48$, this tells us that there will be 5 balloons left over. It's worth doing the maths because if you don't want the kids all to fight and scream about who gets an extra balloon, you know you'll need to discreetly burst five and shove them in the bin before you share the other 48 out.

Now you've got a bucket of 3,721 sweets to share fairly between the 8 kids. Again, the eight times table doesn't go that high so we need to do a bigger division sum. It will look nasty, and when you first saw big divisions at school, you probably blanked out. Never mind, here's another chance to get on board! The secret is that we only need to deal with one little bit at a time and ignore everything else. If you like, you can use a piece of paper to hide the bits you don't need as you go along, and it'll also help you to put all your answers in the right places. We start from the left-hand side and follow this little sequence: 1) divide 2) work out the remainder 3) move along. Ready? We're off . . .

❶ 8|3721

Write the sum out like this. The answer will end up on top.

❷ 8|3 — ignore what's under here

To start with we only have to decide if 8 goes into 3? No, it doesn't, so we need the next digit.

❸ 4 over 8|37 — move this along →

8 into 37 goes 4 times – but what's the remainder?

❹ 4 over 8|37, 32 underneath

Multiply 8 x 4 to get 32 and write it underneath.

❺ 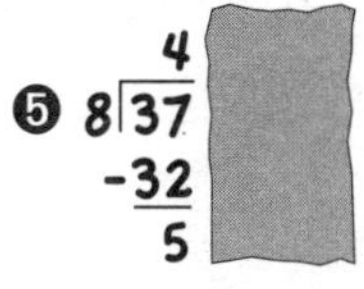

Subtract 37 – 32 to find the remainder is 5. Write the 5 in.

❻ 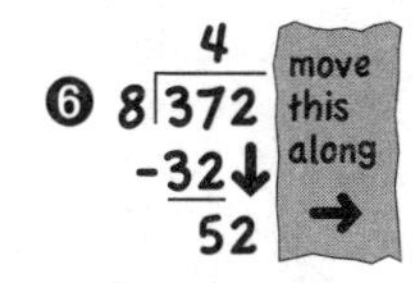

Bring down the 2. Divide 8 into 52.

❼ 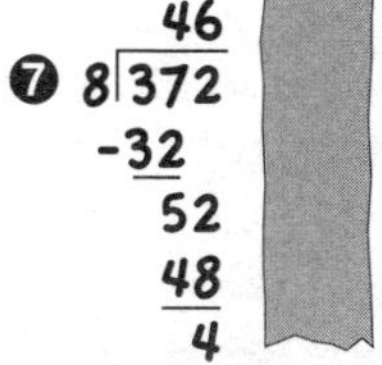

8 into 52 goes 6 times. Write 6 on top then 6 x 8 = 48, so write this underneath the 52 and subtract to get 4.

❽

```
    46
8|3721
 -32
   52
   48↓
    41
```

We've reached the last bit of the sum! Bring down the 1. Divide 8 into 41. It goes 5 times with a remainder of 1.

❾

```
    465
8|3721
 -32
   52
   48
    41
    40
     1
```

8 x 5 = 40 → 40

and

41 – 40 = 1

Write the answer 5 in at the top.

There's nothing else to bring down so the answer is 465 with a remainder of 1.

I've shown every little stage here, but with practice you'll be able to work out things like 37 ÷ 8 = 4 with a remainder of 5 in your head, and you won't need to write out all the fancy stuff underneath. Instead it'll look like this:

Divide the 8 into the numbers starting from the left: 8 into 3 won't go.

8|3721

Move along; 8 into 37 goes 4 times with remainder 5. Write a 4 over the 7 and a little 5 in front of the 2.

$$\begin{array}{r} 4 \\ 8\overline{)37^{5}21} \end{array}$$

Move along; 8 into 52 goes 6 times with remainder 4. Write 6 over the 2 and put a little 4 in front of the 1.

$$\begin{array}{r} 4\ 6 \\ 8\overline{)37^{5}2^{4}1} \end{array}$$

Move along; 8 into 41 goes 5 with remainder 1. Write the 5 in the answer. As we've got to the end the remainder of 1 is the remainder for the whole sum.

$$\begin{array}{r} 4\ 6\ 5 \text{ remainder 1} \\ 8\overline{)37^{5}2^{4}1} \end{array}$$

How to Test if Numbers Divide Exactly by . . .

2 Any even number divides by 2.

3 Add the individual digits. If the answer divides by 3 then so does the number. To test if 438 divides by 3, add 4 + 3 + 8 = 15. As 15 divides by 3, then so does 438.

4 Look at the last two digits. If the 'tens' digit is even, and the last digit is 0, 4 or 8, then it divides by 4. If the 'tens' digit is odd, then the last digit has to be 2 or 6.

5 If your number ends in 5 or 0, then it divides by 5.

6 Because 6 = 2 × 3, if your number is even and also divides by 3 then it will divide by 6.

7 Take away the last digit and multiply it by 2. Subtract the result from the remaining number. If the answer is 0 or divides by 7 then the number you're testing also divides by 7. Let's test 364: take off the 4 and multiply it by 2 to get 8. Subtract 8 from 36 to get 28. As 28 divides by 7, we know that 364 also divides by 7.

9 This works like the test for 3. Add the individual digits. If the total divides by 9 so does the number.

10 This is the simplest test – if your number ends in 0 then it divides by 10!

11 This is really neat. Write out your number putting + and – in front of the digits in turn. You then add or subtract the digits and see if the answer is 0 or divides by 11. To see if 49,137 divides by 11, put in signs: +4 −9 +1 −3 +7 and work out the answer, which comes to 0. This tells you that 49,137 will divide by 11.

Long Division

There are some things in life that you never need to do. You don't need to play golf, you don't need to arrange all your food jars so the labels face out the right way, you don't need to finish the crossword in the newspaper and, thanks to calculators, you don't need to do long division. But . . . if you're secretly wondering if you've got what it takes to wrestle with numbers and beat them, don't be embarrassed. Unlike hobbies such as trainspotting, or synchronized swimming, or shampooing your car, you can do long division in the privacy of your own home and *nobody needs to know*.

It's quite unusual to have to divide one big number by another and get an accurate answer, but occasionally it does happen . . .

In total there are 356 people sharing the money, so to find out how much you get, you need to divide £103,596 by 356. If you've followed this chapter so far, you've already got the general idea of how division works; the only thing that makes long division different is that there's some guessing and bit of extra multiplying involved.

Knocking off Zeros

When we do our guessing it's handy to remember this short cut. Suppose you have 6,000 ÷ 200, you can make the sum much easier by knocking off the same number of zeros from each number. As 200 has two zeros, we can do this: 6,000 ÷ 200 and reduce the sum to 60 ÷ 2 which makes 30. Much easier!

So How Much Money Have You Inherited?

Write the sum out in the same way as before: **356|103596**

Work along the digits from the left until you have a number bigger than 356, like this:

Is 1 bigger than 356? No.
Is 10 bigger than 356? No.
Is 103 bigger than 356? No.
Is 1,035 bigger than 356? Yes!

The first digit of the answer goes here

356|1035

Ignore this bit

This means that the first digit of the answer will go over the 5.

To get this first digit, you need to guess the answer to 1,035 ÷ 356. It helps to round the numbers off roughly so let's say 1,035 is about 1,000 and 356 is about 300. So what's 1,000 ÷ 300? If we knock two zeros off each number, that's the same as 10 ÷ 3 so the answer is 3 and a bit. This suggests that 3 is a good guess, but don't write anything in the answer space yet . . .

We need to test our guess, so we work out 356 × 3 and get 1,068. (This little multiplication can be written to one side out of the way. You can see what I mean on page 35.) The answer has to be less than 1,035, so our guess of 3 is too big. We'll guess 2 instead, and so we work out 356 × 2 and get 712.

Put 712 in underneath the 1,035 and then subtract: 1,035 – 712 = 323. So far, we've worked out that 1,035 divided by 356 goes 2 times with a remainder of 323. So long as the remainder is *less* than 356, our guess of 2 is correct!

356 | 1035
712
323

Now we can write 2 in as the first digit of the answer, and we're allowed to feel a bit smug.

It's time to move along and bring down the next digit, which is 9.

2
356 | 10359
712
3239

Now we need to guess the answer to 3,239 ÷ 356. Let's just dive in and say 8. It might be wrong but if it's right then we'll get a little warm glow of inner satisfaction.

To test it we multiply 356 × 8 = 2,848 and then write that in under the 3,239 and subtract to see what the remainder is.

We find 3,239 – 2,848 = 391. Whoops!

The remainder 391 is bigger than 356, so our guess of 8 was too small. In fact, 356 would go into 3,239 one more time, so 9 should be the right answer.

2
356 | 10359
712
3239
2848
Whoops! 0391

To test it we multiply 356 × 9 = 3,204. We need to rub out 2,848 and write in 3,204, then subtract it from 3,239.

(At this point it's fair to remind you that nobody promised you that this was going to be easy. The reason it's taking so long is to make every little step as clear as possible: with practice all this is much faster.)

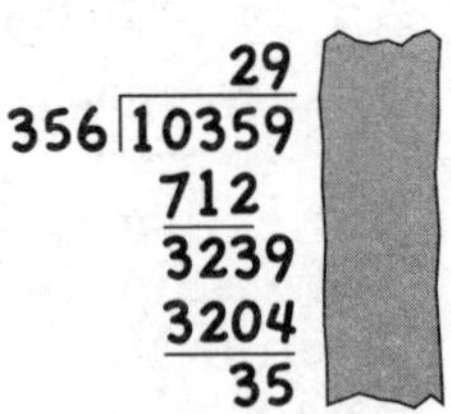

So, we end up with 35 on the bottom of the pile of numbers. This is less than 356, which tells us that the new guess of 9 worked so we plonk it up on the top after the 2 in the next answer space.

Finally, we move along and bring down the last 6.

Now we work out 356 ÷ 356 and we're lucky here because the answer is a nice, neat 1. We can write the 1 in the answer, and if you are that person who has all the labels on your food jars facing the same way, you'll want to tidy things up. You multiply 356 × 1 and write the answer underneath everything else, then you subtract 356 – 356 = 0 to find there's no remainder.

```
       291
356 |103596
      712
      3239
      3204
        356
```

Here's how the completed sum looks, with the little extra multiplications done off to the side. Who knows, you might even have had time to indulge in your artistic side while you were at it.

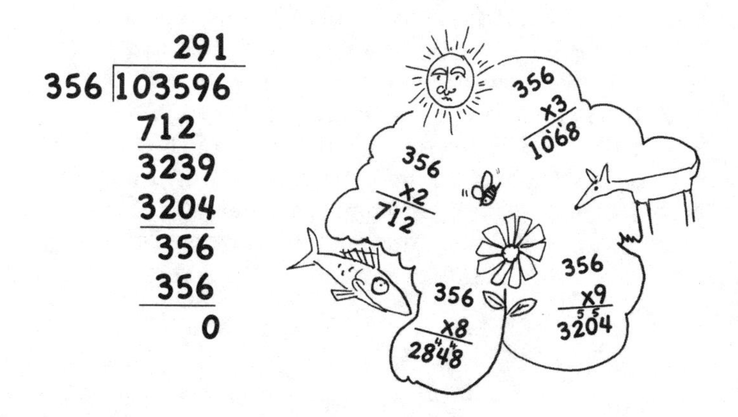

After all the excitement, you might have forgotten that you were working out how much money you'd inherited from your auntie. Well, maybe £291 turned out to be a bit disappointing, but look on the bright side: you've also just acquired 355 cousins to celebrate with!

The Crooked Waiter Problem

Now that we've got past all those big ugly numbers, here's a grand old story about dividing up money which a lot of people know, but hardly anyone can get their head around.

Three ladies go for a light lunch and the bill comes to £30. They each give the waiter a £10 note, but when he gets to the till he realizes that the bill should have been £25. The waiter takes out five £1 coins, but as he walks back to the table he decides to give each lady £1 change and quietly slip the other £2 into his pocket.

So the original bill was £30. The ladies paid 3 × £9 = £27 and the waiter has £2 in his pocket. £27 + £2 = £29. Where did the extra pound go?

It's such a good problem that I don't want you looking straight at the answer; I'll hide it for you to find later on in the book. And don't pull that face: after all, I am giving you all the other answers as we go along!

THE ORDER OF PLAY

When you get a long line of little sums together, it's important to know what order to work them out in. This is the sequence you have to follow:

❶ Bits inside brackets

❷ Powers

❸ Multiply/divide

❹ Add/subtract

Dealing with a Line of Sums

Here's how this sort of thing might arise when the local councillor Mrs Beaumont prepares for a party in the village hall.

Mrs Beaumont went to the bakery and bought two pots of whipping cream for 80p each and three cherry buns for 32p each. What did she spend?

The total = 2 × 80 + 3 × 32. If we just go along the line doing the sums as we reach them we start with 2 × 80 = 160, then 160 + 3 = 163 and finally 163 × 32 = 5,216 or £52·16. That's an awful lot for two pots of cream and three cherry buns! Where did it go wrong?

You must do any multiplying and dividing before adding and subtracting.

Therefore if you have 2 × 80 + 3 × 32, you do the two multiplying sums first, which leaves you with 160 + 96. Finally you do the adding and find that Mrs Beaumont spent 256p, which is £2·56.

Do both bits of multiplying first

$$2 \times 80 + 3 \times 32$$

$$160 + 96$$

Then do the adding

$$256$$

Mrs Beaumont then goes into the party shop because she has to dress up four of her friends. Each lady needs a leopard-print leotard at £17, two cans of spray string at £3 each and a red top hat for £8. As well as all that, Mrs Beaumont wants to buy a fairy wand for herself for £6. As she's such a good customer the shop has given her three £5 vouchers, so how much cash does she need to hand over?

First of all let's see what each of the four ladies needs:

Leotard: £17

Spray string: 2 × £3

Hat: £8

We can write this out as 17 + 2 × 3 + 8. As this is what each lady needs, we'll wrap this up in brackets: (17 + 2 × 3 + 8).

As there are four ladies, we need to multiply everything by 4. We can write it like this 4(17 + 2 × 3 + 8). When you see a number up against the front of brackets it means you need to multiply it by everything inside. Mrs Beaumont also wanted a fairy wand, which makes the total into 4(17 + 2 × 3 + 8) + 6. You'll notice the + 6 is outside the bracket, because we don't want it to be multiplied by 4. Finally she has the three £5 vouchers which are worth (3 × 5). As these are to be subtracted from the total, we need to put a minus sign in front of the bracket. Here's the end result: 4(17 + 2 × 3 +8) + 6 – (3 × 5)

Always work out the bits in brackets first!

Inside the first bracket, we'll do the 2 × 3 multiplication first to get (17 + 6 + 8) and then we can add these numbers to get (31). In the second bracket the (3 × 5) just becomes (15). We've ended up with: 4(31) + 6 – (15).

Remember that the 4 is multiplying the inside of the bracket, so before we get rid of the bracket we have to multiply. This gives us: 124 + 6 – 15 so we finally discover that Mrs Beaumont spent £115.

There's only one question left: if Mrs Beaumont's friends are wearing leotards and red top hats to the village hall party, exactly what is Mrs Beaumont wearing? The answer is that she bought her outfit in the bakery.

MAKING A ROUGH GUESS

When you're doing a sum with big numbers, it's very sensible to have a rough idea of the answer before you start. This especially applies when you're using a calculator because it's so easy to push the wrong button.

The attendance at York City's football match last Saturday was 38,452 and each person paid £27·50 to get in. (These figures are taken from a dream that the manager once had.) The four gate stewards want to know how much money they should have in total so they each use a calculator to work this out: 38,452 × £27·50.

Unfortunately they get four different answers:

a £105,930

b £1,057,430

c £38,479·50

d £105,743,000

Which one do you think is right?

Rounding Off

The first thing to do is make your numbers easier to deal with, so we'll round them off very roughly. You just need to take the first digit and replace the rest with zeros, so 38,452 becomes 30,000. However, to make your guess slightly more accurate, *if the second digit is 5 or more, then add 1 to the first digit.* As the second digit was 8, we should round 38,452 up to 40,000. If you imagine these numbers drawn out on a line like part of a ruler, you can see that 38,452 is closer to 40,000 than 30,000:

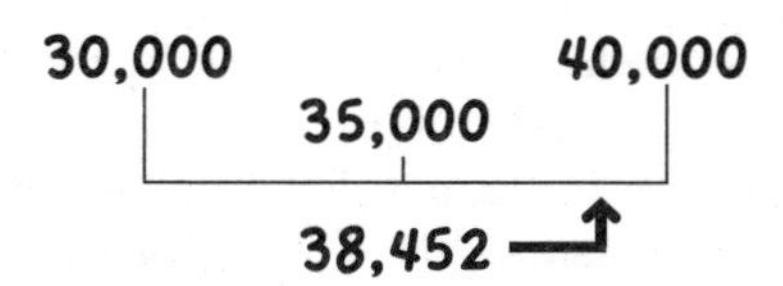

£27·50 could just be rounded to £20, but again the 7 is bigger than 5, so we'll round it up to £30.

There, that'll do. Now work out 40,000 × 30. The sum is just 4 × 3, then you count up the combined zeros and add them on. There are five zeros in total, so the rough answer is 1,200,000. The closest answer to this is b) 1,057,430, so it's the most likely to be right.

Here is where the other three stewards went wrong when they put 38,452 × 27·50 into their calculators: a) the 4 was missed out, c) the + button was pressed instead of the × button and d) the decimal point was missed out from the 27·50.

From this point onwards, some of the trickier sums in this book will have this sign and a brief description of how you might estimate the answer.

FRACTIONS

In the division chapter we used numbers that divided exactly, or we just left the remainders undivided. Things get stranger when you have to divide things up into pieces. You either have to use *vulgar fractions* or *decimal fractions* (also known as just decimals). Sometimes it's easier to use one, and sometimes the other, it depends on what you're doing.

A vulgar fraction (which is usually just known as a 'fraction') is a dividing sum that hasn't been worked out. If you have the sum $4 \div 7$ you can also write it as $\frac{4}{7}$, which is the same as $\frac{4}{7}$ or four sevenths, and the good news is that you haven't had to do any sums. But the bad news is that you've always got two numbers to worry about.

Reducing Fractions

Suppose you've got one pepperoni pizza cut into eight bits. This is the same as $1 \div 8$ so each bit is $\frac{1}{8}$ (or one eighth) of the pizza. If you eat six of them, then you've eaten $\frac{6}{8}$ (or six eighths) of the pizza.

$\frac{6}{8}$ is a perfectly good fraction, but you don't normally see $\frac{6}{8}$ written down and here's why:

$\frac{6}{8}$ = $\frac{3}{4}$ $\qquad$ $\frac{1}{8} + \frac{1}{8} = \frac{1+1}{8} = \frac{2}{8} = \frac{1}{4}$

You can see from the diagram that $\frac{6}{8}$ is the same as $\frac{3}{4}$. This is because two eighths together make one quarter, so six of them make three quarters.

If you don't have any bits of pizza handy to do your sums with, here's how it works with numbers. You start with $\frac{6}{8}$, and then see if you can spot a number that divides into both the top and bottom numbers.

$\frac{6}{8}$ Both numbers will divide by 2 so . . . $\frac{6 \div 2}{8 \div 2} = \frac{3}{4}$ **Ta-dah!**

You can multiply or divide the top and bottom of any fraction by the same number, and the value will still be the same.

It sounds a mess, but all we did here was divide the top and bottom by the same number, 2, and that's how we converted $\frac{6}{8}$ to $\frac{3}{4}$. Suppose we wanted to multiply top and bottom by 6 for some strange reason (you'll see what the strange reason is in a minute), we'd find that $\frac{3}{4} = \frac{18}{24}$.

That's perfectly OK, because if your pizza had been cut into 24 bits and you ate 18 of them, you'd still have eaten the same amount; it's just that the slices would have been a lot thinner.

$\frac{18}{24}$

Making the numbers in fractions smaller is called *reducing* and generally it improves your quality of life. Suppose you've now got a veggie pizza and some maniac has chopped it into 84 pieces. You eat 70 of them, so how much was that? The fraction is $\frac{70}{84}$, which is tricky to get your head round, so once again we're looking for a number that will divide into the top and bottom. As both numbers are even, then we know they'll both divide by 2. That gives us $\frac{35}{42}$ and it turns out that both numbers divide by 7, so now we know you've eaten $\frac{5}{6}$ of the veggie pizza.

Here's the big question: which pizza did you eat more of?

Comparing, Adding and Subtracting Fractions

What we want to know is, which is bigger, $\frac{3}{4}$ or $\frac{5}{6}$?

(The numbers here are quite simple, so it might be obvious to you. You had one quarter of the pepperoni pizza left over, and one sixth of the veggie pizza left over. One sixth is less than one quarter, so you ate more of the veggie pizza.)

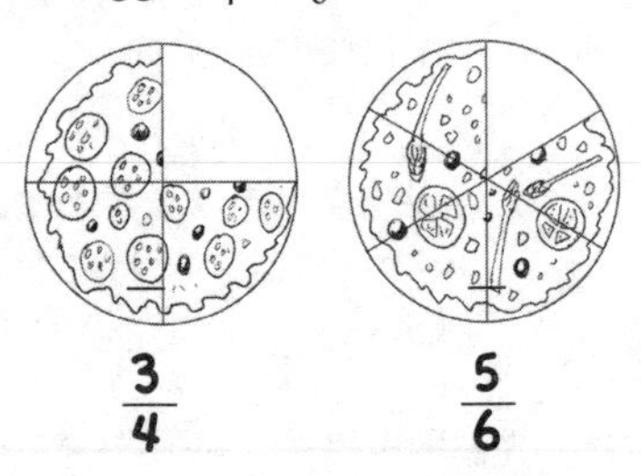

If you want to work this out with sums, you need to make both $\frac{3}{4}$ and $\frac{5}{6}$ into fractions with the same number on the bottom. The most reliable way to do this is to multiply the top *and* bottom of each fraction by the *other* fraction's *bottom*. (If you want to be formal, the top number of a fraction is the *numerator* and the bottom is the *denominator*.)

$\frac{3}{4}$ ← Multiply top and bottom by the bottom of the other fraction ↗ $\frac{5}{6}$ You get: $\frac{3 \times 6}{4 \times 6} = \frac{18}{24}$

$\frac{5}{6}$ ← Do the same with this one ↘ $\frac{3}{4}$ You get: $\frac{5 \times 4}{6 \times 4} = \frac{20}{24}$

So we've found that $\frac{3}{4} = \frac{18}{24}$ and $\frac{5}{6} = \frac{20}{24}$. Therefore $\frac{5}{6}$ is bigger. (You can also compare fractions by converting them into decimals. See page 59.)

The other question you might ask is how much pizza did you eat in total? You ate $\frac{3}{4}$ of the pepperoni and $\frac{5}{6}$ of the veggie, so you need to add $\frac{3}{4} + \frac{5}{6}$.

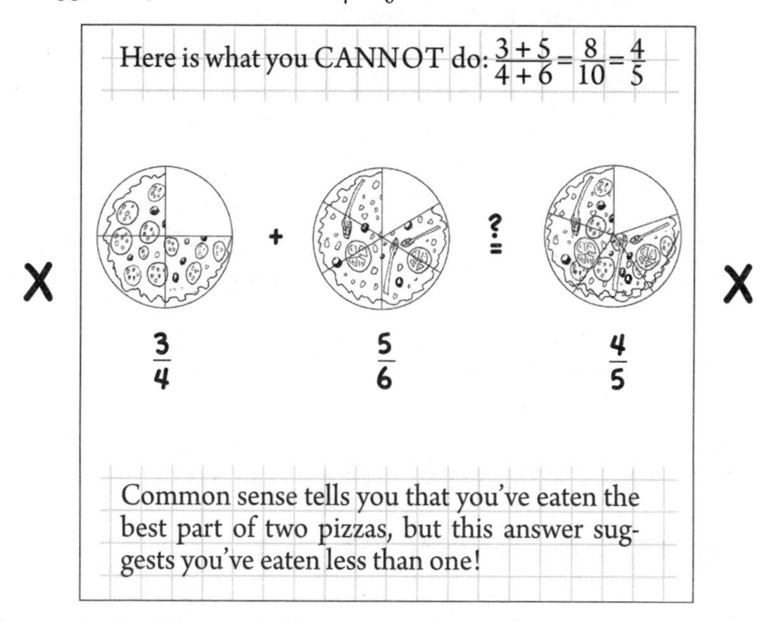

The trouble is that quarters and sixths are not the same thing, so you can't just add them together. You have to convert both fractions so that they have the same number on the bottom. Luckily we did that earlier and found that $\frac{3}{4} = \frac{18}{24}$ and $\frac{5}{6} = \frac{20}{24}$. Once the bottoms are the same we can add the tops together. It's the same as if we'd cut both pizzas up into twenty-fourths.

$$\frac{18}{24} + \frac{20}{24} = \frac{38}{24} = \frac{24}{24} + \frac{14}{24} = 1 + \frac{7}{12}$$

By the time we've finished shuffling bits of pizza about you'll see the $\frac{38}{24}$ turns into 1 and $\frac{7}{12}$. Strange, isn't it? I bet you don't remember eating exactly 7 of anything, and certainly no twelfths, but that's still the answer.

Subtracting fractions works in the same way. Here's $\frac{2}{3} - \frac{3}{5}$ demonstrated once again by our pizzas.

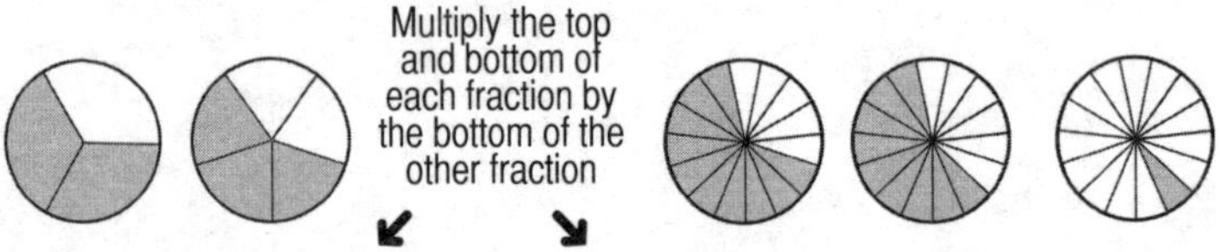

Multiply the top and bottom of each fraction by the bottom of the other fraction

$$\frac{2}{3} - \frac{3}{5} = \frac{2 \times 5}{3 \times 5} - \frac{3 \times 3}{5 \times 3} = \frac{10}{15} - \frac{9}{15} = \frac{1}{15}$$

Mixed Fractions and Improper Fractions

A *mixed fraction* is a whole number with a fraction such as $6\frac{1}{2}$.

An *improper fraction* is a fraction where the top is bigger then the bottom such as $\frac{13}{2}$ (as it happens $6\frac{1}{2} = \frac{13}{2}$).

If you need to convert a mixed fraction into an improper fraction you multiply the whole number by the fraction bottom and then add it to the top. Confused? Then bring on the pizzas again . . .

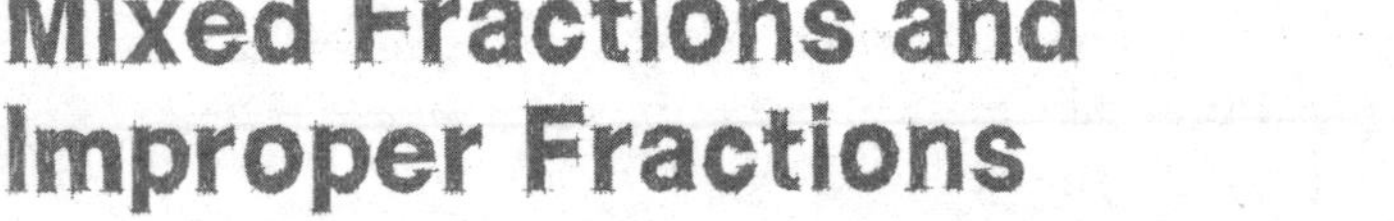

$$2\frac{1}{3} = \frac{2 \times 3}{3} + \frac{1}{3} = \frac{6 + 1}{3} = \frac{7}{3}$$

If you want to convert an improper fraction back into a mixed fraction, you treat it like a dividing sum. To convert $\frac{7}{3}$ you work out $7 \div 3 = 2$ with a remainder of 1. So the answer is 2, and the remainder of 1 stays over the 3, giving you the answer of $2\frac{1}{3}$.

Converting mixed fractions to improper fractions usually makes multiplying and dividing fractions easier, as you're about to find out.

Multiplying by Fractions and the Meaning of 'Of'

Adding and subtracting fractions can be awkward, but thankfully multiplying and dividing isn't so bad.

Multiplying fractions is usually disguised by the word 'of'. If you say 'three quarters of twelve', what you really mean is $\frac{3}{4} \times 12$. When you multiply a whole number by a fraction, there are two sums involved. You multiply by the top and then divide by the bottom. Here's what $\frac{3}{4} \times 12$ looks like:

$$\frac{3}{4} \times 12 = \frac{3 \times 12}{4} = \frac{36}{4} = 9$$

To multiply two fractions, you just multiply the tops together and the bottoms together.

Suppose you spend 7 hours every Saturday and every Sunday birdwatching. What fraction of your week is that? Saturday and Sunday make up $\frac{2}{7}$ of the week, and as there are 24 hours in a day you spend $\frac{7}{24}$ of each day hoping to spot a 'mottled warbler' or a 'crested mudfinch'. Here's the sum:

$$\frac{7}{24} \times \frac{2}{7} = \frac{7 \times 2}{24 \times 7} = \frac{14}{168}$$

At this point you might realize that you can divide both 14 and 168 by 14 and end up with the answer $\frac{1}{12}$, but wouldn't it have been nice to avoid the big numbers in the first place? When you're multiplying fractions, it's always worth seeing if you can reduce them as you go along. The best thing is if you find the same number on the top and bottom because these then *cancel each other out.*

Let's go back to this bit:

$$\frac{7 \times 2}{24 \times 7}$$

Look at the two sevens: the one on top is multiplying everything by 7, but the one on the bottom is dividing everything by 7. If we did both sums we'd end up back where we started, so we needn't bother. Instead we can cross them both out and replace each with a 1. Now that we've done that, let's see what else we can do:

$\frac{{}^{1}\cancel{7} \times 2}{24 \times \cancel{7}^{1}}$ This is the same as: $\frac{2}{24}$ Both the 2 on top and the 24 on the bottom will divide by 2 $\frac{\cancel{2}^{1}}{\cancel{24}^{12}}$ That just leaves . . . $\frac{1}{12}$

So now you know you spend a twelfth of every week bird-watching. That's equivalent to one minute out of every twelve minutes that you're alive, or one solid month every year! (If you start doing the sums for the time you spend on any regular activity such as a hobby or driving to work, it can get quite spooky. For instance, most people spend the equivalent of about 10 days locked in the bathroom every year.)

There's an old maths teaser that turns up in newspapers every once in a while. What is $\frac{9}{10} \times \frac{8}{9} \times \frac{7}{8} \times \frac{6}{7} \times \frac{5}{6} \times \frac{4}{5} \times \frac{3}{4} \times \frac{2}{3} \times \frac{1}{2}$? Whoever prints it probably sits in his office stroking a white cat and laughing like a demon, thinking that everybody will be spending the whole day multiplying and dividing, but of course all the numbers just cancel out until you're left with $\frac{1}{10}$.

Dividing by Fractions

If you divide by a number or a fraction, you just turn the number or fraction upside down and multiply instead.

This sounds mad, but it makes sense when you drive to see your auntie. She lives 10 miles away and you break down half way. How far have you gone? You need to know 'what is half *of* ten' and the *of* tells you that you're multiplying $10 \times \frac{1}{2}$. Of course you get the same answer by asking yourself 'what is 10 divided by 2?'

All whole numbers can be written as fractions with a 1 on the bottom. We can write these two sums out as $\frac{10}{1} \times \frac{1}{2} = \frac{10}{1} \div \frac{2}{1}$. You'll see that dividing by 2 or $\frac{2}{1}$ is the same as 'turning it upside down' and multiplying by $\frac{1}{2}$.

Sums that involve dividing by fractions need thinking about. Suppose you've got a load of chairs to paint, and you know that each chair needs $\frac{2}{3}$ of a tin. (Yes, they are very small tins.) If you have 8 tins, how many chairs can you paint in total?

Quickly think it through. Suppose you were painting tables and they each needed 2 tins of paint, and you had 8 tins. How many tables could you paint? The sum is $8 \div 2 = 4$ tables.

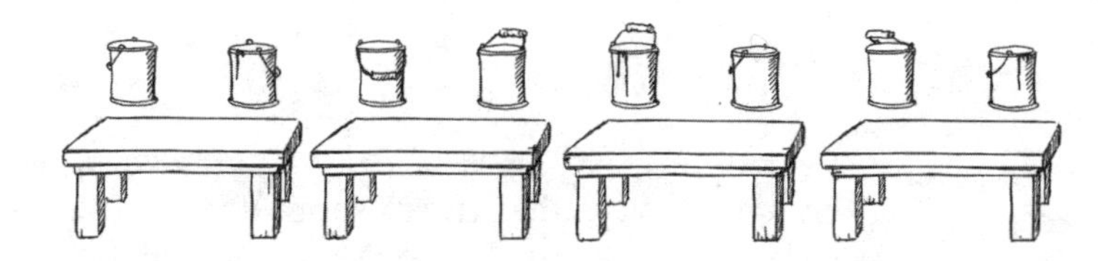

The important bit is that you were dividing by the number of tins each table needed. The same happens with the chairs: you divide by the number of tins each chair needs. So the number of chairs you can paint is $8 \div \frac{2}{3}$. To work this out, you turn the $\frac{2}{3}$ upside down and multiply:

$$8 \div \frac{2}{3} = 8 \times \frac{3}{2} = \frac{8 \times 3}{2} = \frac{24}{2} = 12$$

So 8 tins will cover 12 chairs. Notice that when you divide by a fraction, you end up with a bigger number than the one you started with.

It's always worth checking answers like this to see if it makes sense. If each chair needs $\frac{2}{3}$ tin, then one tin can paint a chair and you'll have some paint left over. Therefore 8 tins will be able to paint 8 chairs plus a few more, so our answer of 12 looks pretty good.

A Fishy Problem

Before we leave fractions let's get a really horrible sum and put it to bed, just for the sheer masochistic hell of it. How many oil drums are needed to fill this fish pond?

Your fish pond needs $85\frac{1}{2}$ buckets of water

One oil drum holds $11\frac{1}{4}$ buckets

$\times\ 11\frac{1}{4} =$

Before you start – *have a rough guess!* It gives you an idea of what the answer should be, and also helps you set the sum up correctly. The oil drum can carry about the same as 10 buckets. The pond needs about 80 buckets. So the number of oil drums you need will be about $80 \div 10 = 8$. That seems reasonable.

Now we need to use the exact numbers. Instead of $80 \div 10$, we'll work out $85\frac{1}{2} \div 11\frac{1}{4}$.

The first thing to do is turn the mixed fractions into improper fractions: $85\frac{1}{2} = \frac{85 \times 2 + 1}{2} = \frac{171}{2}$ and $11\frac{1}{4} = \frac{11 \times 4 + 1}{4} = \frac{45}{4}$. Now we just need to work out $\frac{171}{2} \div \frac{45}{4}$. Here we go . . .

$$\frac{171}{2} \div \frac{45}{4} = \frac{171}{2} \times \frac{4}{45} = \frac{171}{\cancel{2}_1} \times \frac{\cancel{4}^2}{45} = \frac{\cancel{171}^{19}}{1} \times \frac{2}{\cancel{45}_5} = \frac{19 \times 2}{5} = \frac{38}{5} = 7\frac{3}{5}$$

Turn upside down and multiply

CANCEL! Top and bottom will divide by 2

CANCEL EVEN MORE! Top and bottom also divide by 9

Sorted!

We've got an answer of $7\frac{3}{5}$ oil drums which is quite close to our rough guess of 8, so we can feel confident. We can also feel happy, victorious and unbearably pleased with ourselves.

RATIOS

Ratios crop up every day – whether you're diluting a bottle of floor cleaner, doubling up the ingredients in a recipe or simply trying to get the picture on your TV to look right. You need to understand them if you don't want to ruin your floor, poison your friends or make weather forecasters look like sumo wrestlers . . .

How Big Is Your TV?

A ratio describes the shape of your TV screen, using two numbers to compare the width to the height. Traditional TV screens have an aspect ratio of 4:3 which means that if the screen was 400mm wide, it would be 300mm high. In the old days, if you got a bigger telly, you needed the same shaped screen; otherwise people would look too thin or too fat. This meant that the aspect ratio had to be the same. If your bigger screen was 600mm across, how high should it be? You can make the ratio into a fraction of either $\frac{3}{4}$ or $\frac{4}{3}$ and multiply, but which fraction do you use? It's time for common sense. You want the height to be less than the width, so you multiply 600mm × $\frac{3}{4}$ = 450mm. Although the screen is bigger, it will be the same shape.

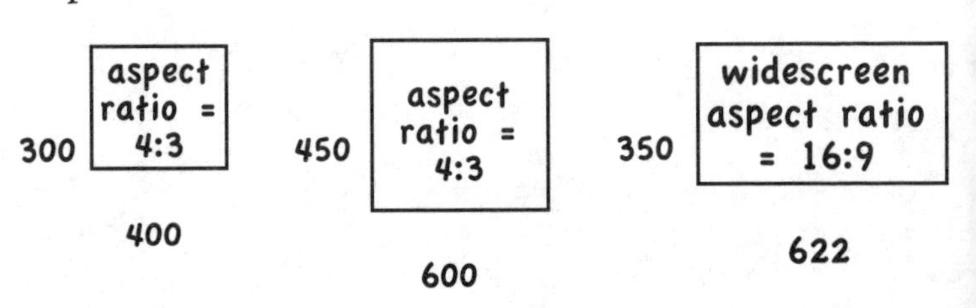

New widescreen TVs usually have an aspect ratio of 16:9 so if your screen was 350mm high, how wide will it be? It has to be wider than high, so you multiply $350 \times \frac{16}{9}$ and get the answer: 622mm.

That's how a bit of simple maths keeps your TV picture looking natural, whatever size of TV you have. Unfortunately maths can't improve the programmes.

The Shadow Stick

You can use ratios to work out the height of a tall statue (or tree or building). You need a sunny day, a stick and a tape measure. Put the stick straight up in the ground, then measure its height, the length of its shadow and also the shadow of the statue.

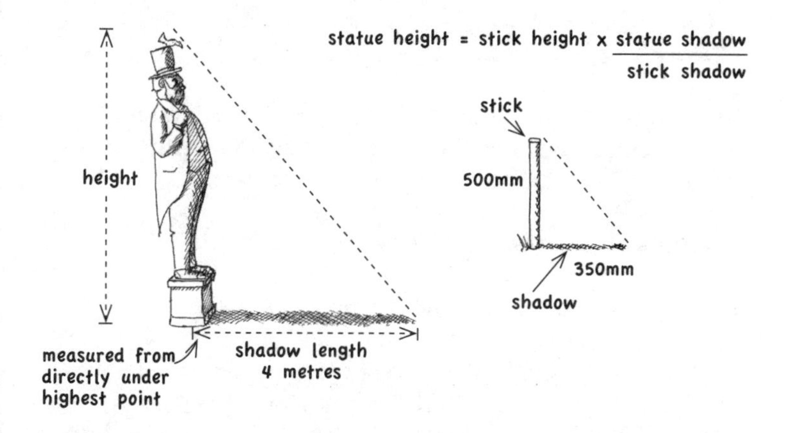

The sunlight creates two *similar* triangles, which means that they are different sizes but exactly the same shape. The ratio of the stick height to its shadow will be the same as the ratio of the statue height to the statue shadow.

The stick measures 500mm high and the shadow is 350mm long. We know the ratio of height to shadow is 500:350. Once we've got the ratio, we're allowed to reduce it just like a normal fraction:

500:350 → Both numbers will divide by 10 → **50:35** → Now both numbers will divide by 5 → **10:7** That's better!

So we've found out that the ratio of height:shadow is 10:7. If the statue's shadow is 4 metres long, we can now work out the height of the statue. The stick height is bigger than the stick shadow, so the statue height must be bigger than 4 metres. Therefore we multiply the length of the statue's shadow by $\frac{10}{7}$ and find that the statue height is $\frac{40}{7}$, or $5\frac{5}{7}$ metres.

Mixture Ratios

If you want to mix up some concrete, you need to know how much cement powder, sand and aggregate (crushed stone) you need. A typical mix might have the ratio as:

cement:sand:aggregate = 1:2:4

If you knew you had 3 tonnes of sand and wanted to use it all, how much cement and aggregate would you need?

The ratio tells us that for every tonne of cement you would need 2 tonnes of sand and 4 tonnes of aggregate. You can alter the numbers in a ratio by multiplying them all by the same thing. Here we want to change the 2 tonnes of sand to 3 tonnes, so we multiply all three numbers in the ratio by $\frac{3}{2}$. The ratio becomes $\frac{3}{2}$:3:6.

This tells us that to make concrete with 3 tonnes of sand we would need to add $1\frac{1}{2}$ tonnes of cement and 6 tonnes of aggregate.

If you wanted to make 10 tonnes of concrete, how much of each ingredient would you need? From 1 tonne of cement, 2 tonnes of sand and 4 tonnes of aggregate you would get 7 tonnes of concrete.

To make 10 tonnes of concrete you multiply each number by $\frac{10}{7}$. You'd need $\frac{10}{7}$ tonnes of cement, $\frac{20}{7}$ tonnes of sand and $\frac{40}{7}$ tonnes of aggregate, which is the same as $1\frac{3}{7}$ tonnes of cement, $2\frac{6}{7}$ tonnes of sand and $5\frac{5}{7}$ tonnes of aggregate.

DECIMALS

Vulgar fractions such as halves and thirds are usually the most obvious way to divide up chunky things like pizzas and tins of paint. However, when you're dealing with numerical things like measurements and money, working with decimal fractions can make life easier. (We usually just call them 'decimals' for short.) Also, if you have two very different looking vulgar fractions such as $\frac{14}{19}$ and $\frac{27}{35}$, it's much easier to compare them if you convert them to decimals, as you'll see later on.

What Happens Beyond the Decimal Point?

The mixed fraction $731\frac{5}{8}$ can also be written as a decimal: 731·625. Obviously the 731 bit is the same for both numbers, the tricky bit is that $\frac{5}{8}$ = 0·625. We'll see how to convert $\frac{5}{8}$ to a decimal soon but first let's see what all the different digits represent.

7	3	1	•	6	2	5
100s	10s	1s	DECIMAL POINT	$\frac{1}{10}$	$\frac{1}{100}$	$\frac{1}{1000}$
Hundreds	Tens	Units		Tenths	Hundredths	Thousandths

The 7 is worth 7 hundred, the 3 is worth 3 tens, and the 1 is just worth 1 unit. As you move along the line of digits, each digit is worth ten times less; once you get past the decimal point, the 6 is worth 6 tenths, the 2 is two hundredths and the 5 is 5 thousandths. Decimals can go on into *tenths of thousandths, hundredths of thousandths,* and so on, but it gets very fiddly to calculate, and almost as difficult to pronounce. Go on, read the last sentence out loud, especially if you're eating cream crackers on the bus.

When we write decimals, if there's nothing in front of the point we put a zero just to make it clear there's a point there.

Rounding Off Decimals

If you're converting fractions to decimals, some of them will work out exactly to a few decimal places, but others form an endless chain of digits going into millionths and billionths, so you need to decide how accurate you want to be and then round the decimal off. For example, the exact decimal for $\frac{1}{6}$ is 0·166666666 . . . and the sixes go on for ever. Life isn't long enough for this, so if we round it to three decimal places, the exact value of $\frac{1}{6}$ is somewhere between 0·166 and 0·167. To see which is closer, you look at what the next digit would have been, which of course is another 6. *If the digit is 5 or more you round upwards.* This means you go up to 0·167. As we saw earlier, if you're not sure, you might find it easier to imagine these numbers as marks on a ruler.

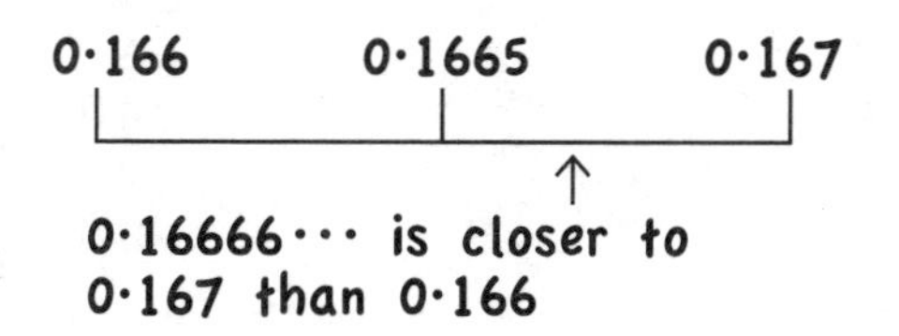

If you divide one number by another and don't get an exact answer, then the decimal digits will always repeat eventually. These decimals are called *rational*.

Converting Between Fractions and Decimals

As we've seen before, normal fractions such as $\frac{5}{8}$ are just dividing sums that haven't been worked out. A decimal is what you get when you do work it out.

From Fractions to Decimals

If we try the sum 5 ÷ 8, we start off trying to divide 8 into 5, which won't go. (If we were allowed to have remainders we'd say 8 into 5 goes 0 times with a remainder of 5.) To work the answer out as a decimal, imagine your number 5 is written as 5·000000. We then do a normal division, and when we get past the decimal point we start bringing down the zeros.

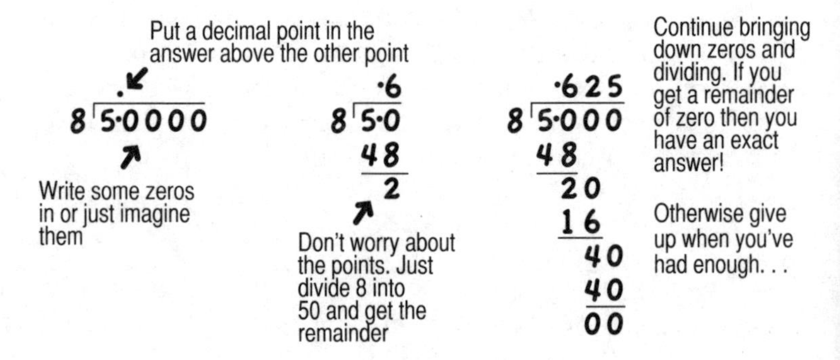

And there's the answer: $\frac{5}{8} = 0{\cdot}625$.

Of course we could have got the same answer by putting 5 ÷ 8 into a calculator. I just showed you the division sum so that you can see for yourself exactly what's going on.

From Decimals to Fractions

If the decimal is just one digit long then the fraction will be tenths so $0{\cdot}6 = \frac{6}{10}$ and this one will then reduce to $\frac{3}{5}$.

If the decimal is two digits long then the fraction will be hundredths so $0{\cdot}75 = \frac{75}{100}$ which reduces to $\frac{3}{4}$. However 0·76 will only reduce to $\frac{19}{25}$ and 0·77 is $\frac{77}{100}$ and it won't reduce at all.

Almost any other decimal can be nasty to convert. If you were faced with 0·692308, the best thing is to round it off to 0·7 and then say it's about $\frac{7}{10}$. (The actual answer is that $0{\cdot}692308 = \frac{9}{13}$ rounded to six decimal places but you'd given up caring hadn't you?)

How Decimals Can Help with Fractions

As we've already seen, adding and subtracting fractions can be a bit tedious, but if you convert them to decimals, it's much easier. Remember on page 44 when we added $\frac{3}{4}$ to $\frac{5}{6}$ to see how much pizza we'd eaten?

A calculator tells us that $\frac{3}{4} = 0{\cdot}75$ and $\frac{5}{6} = 0{\cdot}83333$. Therefore $\frac{3}{4} + \frac{5}{6} = 1{\cdot}58333$. Using a calculator and decimals is certainly faster and easier, but it does leave you wondering what 1·58333 of a pizza looks like.

A calculator can also help you to compare fractions. Which is biggest: $\frac{14}{19}$ $\frac{27}{35}$ $\frac{32}{41}$ or $\frac{36}{47}$? Convert them to decimals, and the answer is obvious! In the same order, they become 0·737, 0·771, 0·780 and 0·766. The biggest decimal is 0·780, so $\frac{32}{41}$ is the biggest fraction.

Some Curious Decimals

- $\frac{1}{9}$ = 0·1111111 . . . $\frac{2}{9}$ = 0·2222222 . . . $\frac{3}{9}$ = 0·3333333 . . . and so on.
- $\frac{1}{11}$ = 0·090909 . . .
- $\frac{1}{7}$ = 0·142857142857142857 . . . and the same cycle of digits turns up in $\frac{2}{7}$, $\frac{3}{7}$, $\frac{4}{7}$, $\frac{5}{7}$ and $\frac{6}{7}$. So $\frac{2}{7}$ = 0·2857142857142857.
- $\frac{1}{9801}$ = 0· 00 01 02 03 04 05 06 07 08 09 10 11 12 13 . . . and so on.

Multiplying and Dividing by 10,100 or 1,000

To multiply whole numbers by 10 you just plonk a zero on the end, so 37 × 10 = 370. However, it is more useful and accurate to imagine you're moving the digits along one column *to the left.*

If we want to *divide* by 10, we move the digits one column *to the right,* so 37 ÷ 10 = 3·7. This time the digits move across the decimal point so we need to write it in. If we divide by 100 we move two places, and if we divide by 1,000 we move three places. When gaps appear between the digits and the decimal point we have to fill them in with zeros, so 37 ÷ 1,000 = 0·037.

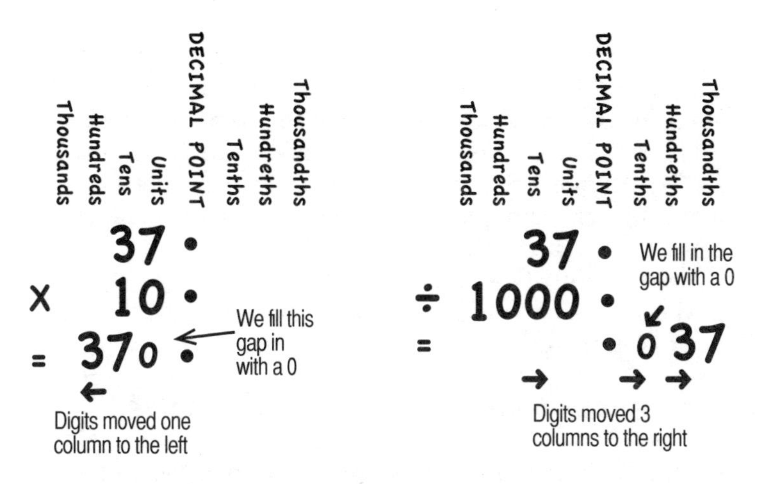

You can multiply and divide decimal fractions just as easily as whole numbers by moving them to the left and right: 0·0451 × 100 = 4·51 or 0·0023 ÷ 10 = 0·00023

Sums with Decimals

Adding and subtracting decimals is quite easy. You line your numbers up as if doing a normal sum; just make sure the decimal points are all in line. If you want to work out 4·07 – 0·256, it looks like there's nothing to subtract the 6 from. Don't panic. You can just put an extra zero on the end of the 4·07 so the 6 isn't feeling too lonely.

$$\begin{array}{r} 4{\cdot}070 \\ -\,0{\cdot}256 \\ \hline =3{\cdot}814 \end{array}$$

However, unless you're sitting an arithmetic exam, or helping a twelve-year-old with their homework, you're very unlikely ever to need to multiply (or divide) decimals without a calculator. But just in case you do find yourself stuck in a calculator-less situation . . . Mrs Beaumont is a fan of Dairylight yoghurts because the fraction of fat they contain is only 0·04. It doesn't sound much, but if she swallows 1·2 litres of the stuff, how much fat has she eaten?

The sum is 1·2 × 0·04. If you want to multiply some humble little decimals like this, first count up how many digits come after the points. Here we've got three digits after the points (the 2, the 0 and the 4). Now just multiply the numbers without the points. 12 × 04 = 48. Now put in the decimal point, making sure the same number of digits come after it. Here we need three digits after the point so we get the answer 0·048.

Mrs Beaumont has swallowed 0·048 of a litre (or 48 millilitres) of fat, which would be enough to make a candle about the size of a carrot. Yuck.

What About More Complicated Decimals?

Decimals are usually a lot nastier than this when you find yourself doing conversions. There's a whole section about how to change litres to pints and metres to inches on page 117 but for now here are just a couple of examples to show what to do with the numbers. It always helps to start with a rough guess first.

You're flying to a line-dancing convention, and the luggage limit is 23kg. Your trusty old scales tell you that your case weighs 48·1 pounds, so are you allowed to take it? First you need to know that 1 pound weight = 0·454kg. This means that the weight of your case in kg will be 48·1 × 0·454.

48·1 is around 50 and 0·454 is nearly 0·5, so the answer should be about 50 × 0·5 = 25.

Oh dear! Your rough guess of 25kg looks like your case might be too heavy, but before you leave behind your beloved monogrammed cowboy boots, let's find the accurate answer. You might prefer to write it out as fractions so that you get whole numbers on the top.

$$48{\cdot}1 \times 0{\cdot}454 = \frac{481}{10} \times \frac{454}{1000} = \frac{481 \times 454}{10 \times 1000} = \frac{218374}{10000} = 21{\cdot}8374$$

Thank goodness, your case comes in at just under 22kg so you can take your boots after all. Yee-hah!

You can also multiply decimals using the grid method on page 22.

You can divide by decimals in the same way. Suppose you're in a charity shop and see some fabulous orange disco trousers with a waist size of 32 inches. The assistant measures you and rather unhelpfully says that your waist is 1·14 metres. If 1 inch = 0·0254 metres, is it going to be horrendously embarrassing if you try them on? The sum to find your waist in inches is 1·14 ÷ 0·0254.

1·14 is close to 1 and 0·0254 is close to 0·03 or $\frac{3}{100}$ so a rough answer is $1 \div \frac{3}{100}$. Dividing by $\frac{3}{100}$ is the same as multiplying by $\frac{100}{3}$. (See 'Dividing by Fractions' on page 49), so our rough answer becomes $1 \times \frac{100}{3} = \frac{100}{3}$ = about 33.

It looks like the trousers might be worth risking, but to be on the safe side, we'll work out your waist size exactly:

$$1{\cdot}14 \div 0{\cdot}0254 = \frac{114}{100} \div \frac{254}{10000} = \frac{114}{100} \times \frac{10000}{254} = \frac{1140000}{25400} = \frac{11400}{254} = 44{\cdot}88$$

The answer is that your waist is 44·88 inches, so the orange trousers are likely to explode in the fitting room, but don't take it too badly. It would be less embarrassing than actually being seen dancing in them.

You might think that 44·88 was a long way off the rough guess of 33, but with all those zeros flying around, the guess is mainly to make sure you've got the decimal point in the right place. If you'd got an answer of 4,488 inches or 0·04488 inches, then it's time to be worried!

POWERS AND ROOTS

Most of us will never need to use powers or roots, although they do creep in if you're dealing with areas and volumes, which are explained later in the book. However, if you happen to design racing cars or plan space travel, they are essential for working out speed, accelerations, braking distances and fuel consumption.

Squares and Square Roots

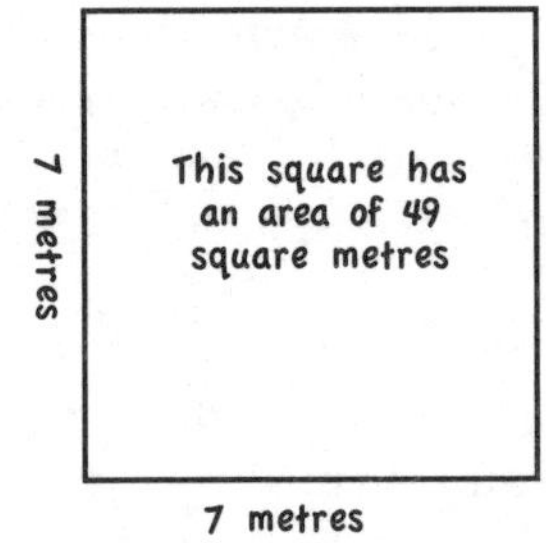

We've already seen how square numbers appear on the multiplication tables on page 15. Squared numbers are usually associated with area calculations and there are lots of ways of describing them: 7 squared is the same as 7×7 . . . which we can write as 7^2 . . . which can also be called 7 to the *power* of 2 . . . and however you write it or whatever you call it, it comes to 49.

7 metres

This square has an area of 49 square metres

7 metres

Suppose you start with 49 and want to work backwards: you need to work out what number multiplied by itself makes 49. This is called the *square root* of 49, which can be written as

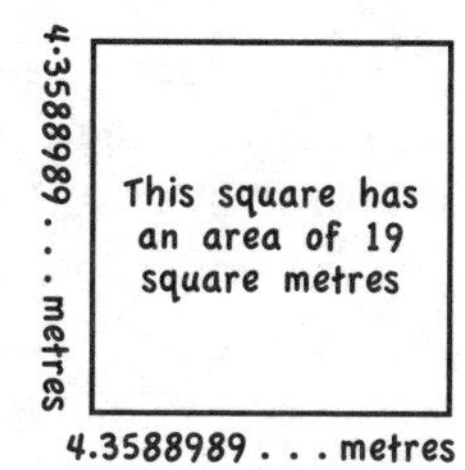

$\sqrt{49}$ . . . or even as $49^{\frac{1}{2}}$. . . which is called 49 to the *power* of a half . . . but however you write it or whatever you call it, it comes to 7. (The square root of 49 can also be –7 because multiplying two negatives makes a positive.)

Getting the square roots of square numbers such as 1, 4, 9, 16 and 25 is easy because the roots are whole numbers. Getting the square roots of other numbers is much nastier. For instance, 19 is not a square number, so if you had a square with an area of 19 square metres, how long would each side be?

The answer is $\sqrt{19}$, but exactly how big is that? We know that $\sqrt{16} = 4$ and $\sqrt{25} = 5$ therefore the square root of 19 will be somewhere in between 4 and 5.

Getting an exact answer with a pencil and paper requires a bit of mental gymnastics, so you're forgiven for grabbing a calculator. You just push <19 √> and get the answer 4·3588989 . . . and this is a decimal that goes on forever without repeating or making a pattern. Decimals that do this are called *irrational.* All square roots that are not whole numbers are irrational.

Other Powers and Roots

You can have any other power you like, but the only one you're ever likely to see are cubes, e.g. 6^3 = *six to the power of three* = $6 \times 6 \times 6 = 216$. Cubes are mainly used in volume calculations, and the very simplest would involve a cubic tank (i.e. a tank with square sides, top and bottom).

If you need to reverse the cubing process it's called a *cube root* and the sign looks like a square root but a little^3 has appeared by it: $\sqrt[3]{216} = 6$.

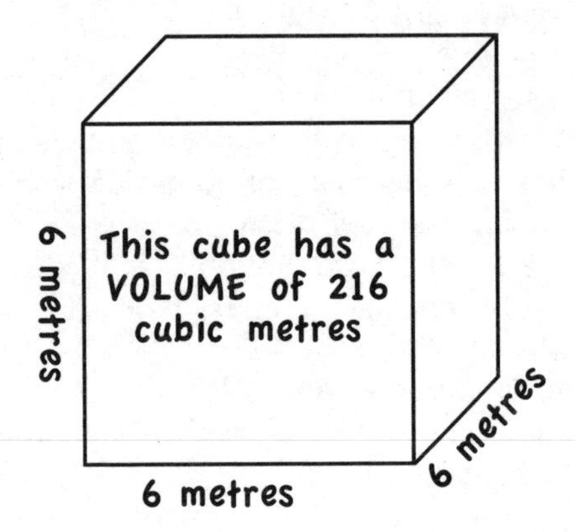

Therefore if you knew that a tank had a volume of 216 cubic metres, then the length of each side is the cube root of 216, which is 6 metres.

If you have a *negative* power then you divide by the number. For instance 10^{-3} is ten to the power of minus three. This is the same as:

$$\frac{1}{10^3} = \frac{1}{10 \times 10 \times 10} = \frac{1}{1000} = 0{\cdot}001$$

Negative powers turn up when you're dealing with massively big or ridiculously small numbers, as you'll see in the next section.

Standard Form

The Earth's mass is about 6,000,000,000,000,000,000,000,000kg

The posh name for this number is six *septillion*, but it's much easier to say 'six with twenty-four zeros on the end'. You can describe it in the same way using numbers, like this:

The Earth's mass is about 6×10^{24}kg

Suppose we were multiplying 6×10^3. This is the same as $6 \times 1,000$, so we would move the 6 three places to the left and get 6,000. In the same way 6×10^{24} means 6 with 24 zeros on the end.

If you wanted to make the weight more accurate, instead of just having a single digit, such as 6, you'd use a decimal with just one number before the point and then multiply by the power of ten like this:

The Earth's mass is $5{\cdot}9736 \times 10^{24}$kg

Writing numbers out like this is called *standard form*. Even though there are a lot more digits in the decimal, you'll see that the $\times 10^{24}$ hasn't changed. When you multiply by 10^{24} you still move the digits twenty-four places to the left and fill in the gaps with as many zeros as you need after the decimal digits. As the digits 9736 already take up four places, we just add on 20 zeros to get the weight as 5,973,600,000,000,000,000,000,000kg.

You can also use standard form for very small numbers:

The mass of one hydrogen atom is $1{\cdot}67 \times 10^{-27}$kg

At first glance it looks like a hydrogen atom has a bigger mass than earth. It's the tiny minus sign that makes all the difference because, instead of multiplying, you divide. Therefore $\times\ 10^{-27}$ is the same as $\div\ 10^{27}$ and so you need to move everything twenty-seven places to the right.

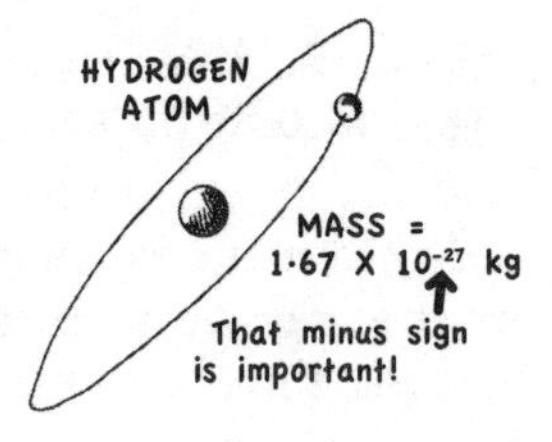

The mass of one hydrogen atom is 0·00000000000000000000000000167kg

Most calculators use this system to show numbers that they can't fit on the screen. Instead of $1{\cdot}67 \times 10^{-27}$ they are more likely to show it as 1·67 E –27. (The 'E' means 'exponential' or 'power'.)

AVERAGES

Averages crop up a lot in the news, especially when they want to give you shock statistics such as the rise in average global temperatures, or the average amount of sleep teenagers need, or the average number of cars footballers own. There are actually three sorts of averages – mean, mode and median – but when people talk about 'the average' they usually mean the mean.

The Mean

Working out averages can help you predict what's going to happen. For instance, if you had 7 days on holiday in Brownpool last year and spent £350 in total, the mean cost per day was £350 ÷ 7 = £50. If you're planning to go for 10 days this year, you know you'll need about £50 × 10 = £500. Now we'll look at how means can help an enterprising businessman...

Last Saturday Lardy parked up his pie van right next to the fence around a health farm. Forty people reached through the barbed wire to buy Lardy's pies. One person just bought one pie, fifteen people bought two pies and so on. The other results are here:

Pies bought per person	1	2	3	4	5	6	7
Number of people	1	15	9	4	6	3	2
Total pies (pies x people)	1	30	27	16	30	18	14

To work out the mean number of pies that each person bought you need to work out the following:

total number of pies sold ÷ total number of people

To find the total number of pies sold you add all the numbers on the bottom row together, which comes to 136 pies, and if you add up the numbers in the middle row, you find the total number of people, which is 40.

136 ÷ 40 = 3·4

Therefore the mean number of pies bought by each person was 3·4. Now that Lardy knows the mean number, he can work out how many pies to order in future. Suppose he expects to serve 1,000 people before he's finally arrested, the total number of pies he can expect to sell will be about 3·4 × 1,000 = 3,400.

The Mode and the Median

The mode is the number that turns up most often in a group of results. More people bought 2 pies than any other quantity, so 2 is the mode. If you stopped a person at random as they waddled back to the shower block and asked how many pies they'd had, the most likely answer would be two.

The median is the value in the middle of the range of results, and can be useful as a quick rough guide to the mean value. If we have an *odd* number of results, it's very easy to pick out. Lardy notes down the ages of his first five customers, then arranges them in order:

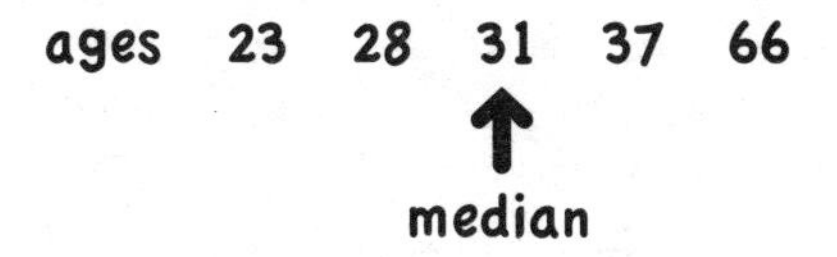

Here the median is 31.

If you have an *even* number of results then you find the two in the middle and work out their mean. Lardy has found out the weight of eight loyal customers and puts them in order:

weight in kg 61 72 72 73 78 84 89 112

median is the mean of 73 and 78

$73 + 78 = 151$ then $\frac{151}{2} = 75{\cdot}5$

The middle two values are 73 and 78, so the the median weight of his eight loyal customers is 75·5kg.

ALGEBRA

When you were learning maths at school, even if you survived long division and decimals, algebra is probably the point at which you ran screaming towards the fire exit. It's quite understandable. Doing sums with a few numbers makes sense, because that's what numbers are for. But doing sums with letters? It *can't* make sense . . . or can it?

What's the Point?

Algebra works like a language. It's a quick way to describe and lay out problems and, once you've mastered the basic rules, it gets the answers with minimum fuss. It's like being abroad and wanting to ask somebody if they know where there's a vegetarian restaurant. You can spend hours trying to mime eating a nut roast and parsnips, but if you know the language you can get the answer in a few seconds.

As for using letters, they just represent numbers that you don't know yet. Later on, we've got a sum to work out the price of a cup of coffee, but because it's a bit tedious writing out 'the price of a cup of coffee' every time we just call it c for short.

We're not even going to bother with letters to start with, we'll just see how a few numbers work on their own, and this will make the basic rules obvious.

Positive, Negative and Equals Signs

This little sum has an equals sign in the middle, so it's called an *equation*:

$$7 - 2 = 4 + 1$$

The numbers on the left-hand side (or LHS) make 5 and the numbers on the right-hand side (or RHS) also make 5. The key to algebra is knowing how to rearrange the numbers and letters in an equation so that you can work out the answer.

Every number has to be either positive or negative.

All negative numbers always have a – sign in front of them. All positive numbers should have a + sign, but we don't always bother writing it in.

It's fun to think of an equation as a see-saw with the equals sign as the pivot in the middle. The positive numbers are weights pushing down and the negative numbers are balloons pulling upwards.

$$7 - 2 = 4 + 1$$

If you want to move the numbers around on one side, their signs have to move with them. If we swap the numbers on the LHS we get:

You'll see the minus sign has to stick with the 2 or the equation wouldn't be right. In the equation the 7 has gained a + sign to remind us that it's positive. Now suppose we want to get the +7 on its own on the LHS. There is just one golden rule:

You can do whatever you like* to equations, so long as you treat both sides exactly the same.

To get the +7 on its own, we need to get rid of the −2 on the LHS. The way to do it is to add +2, but we must follow the rule and add +2 to both sides of the equation.

* The only thing you can't do is divide both sides by zero, because it makes the universe collapse. You'll see why on page 86.

The −2 and the +2 on the LHS make 0, so they cancel each other out. We still have the new +2 on the RHS, so we get:

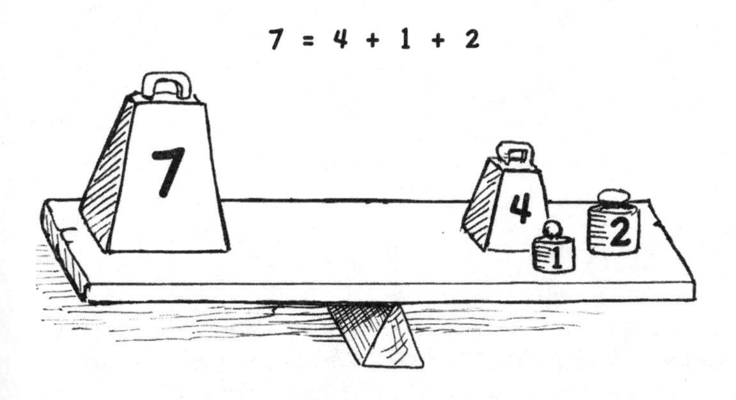

If you check the numbers you'll see that 7 does equal 4 + 1 + 2, and meanwhile we've demonstrated a little trick:

We started with this... and finished with this.

$-2 + 7 = 4 + 1$ $\qquad$ $+7 = 4 + 1 + 2$

The -2 swapped sides and turned into +2

If you move a number across the = you change the sign! Therefore − becomes +, and + becomes −.

Here's another thing that the see-saw can show us: you can swap the two sides over:

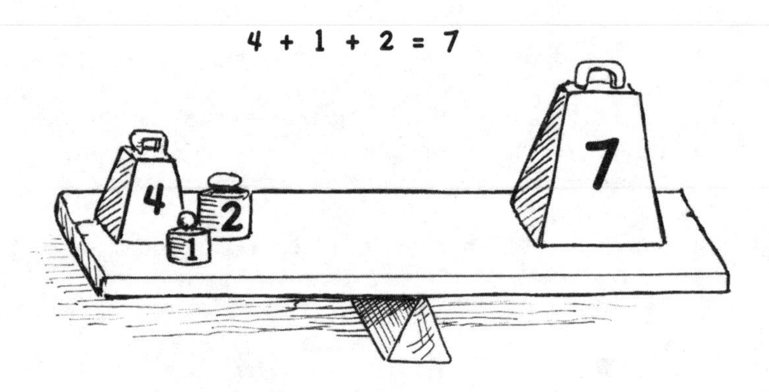

Brackets

Let's keep with $7 = 4 + 1 + 2$ for a moment. Suppose we wanted to know what 14 equals? We need to multiply the 7 by 2, so we must do this to both sides of the equation. There are three bits on the RHS and all of these must be multipled by 2. Here's how it looks:

2 X 7 = 2(4 + 1 + 2)

You'll see we've put everything on the RHS into a bracket. We could have written it out like this: $2 \times 4 + 2 \times 1 + 2 \times 2$ but the brackets save time. The 2 at the front of the bracket is called a *coefficient.*

If you have a number leaning against the front of a bracket, you multiply it by everything inside.

Bring on the Letters

By now you'll be desperate to get stuck into some big fat simultaneous differential equations, but it's worth taking it slowly at first.

Out on the street you bump into Malcolm who is looking a bit shocked. He's just taken his mum into Barstucks coffee shop with £10. He bought two coffees and now has just £1·20 left. What did each coffee cost? Here's what we know:

£10 minus the cost of 2 coffees = £1·20

We can save a lot of ink if we say that the cost of one coffee is c. This means that the cost of two coffees is $2 \times c$ but we usually just write this as $2c$.

Let's make it into an equation and see how we get on:

$$£10 - 2c = £1{\cdot}20$$

We want to jiggle this round so that we have $c =$ by itself on the left. The first thing we'll do is move the £10 across and change its sign to minus:

$$-2c = £1{\cdot}20 - £10$$

That minus sign in front of the $2c$ is a bit ugly, so we'll get rid of it. We multiply everything on both sides of the equation by -1. This changes any + sign to – and changes any – sign to +:

$$2c = £10 - £1{\cdot}20$$

Now we work out £10 – £1·20 = £8·80:

$$2c = £8{\cdot}80$$

We just want c by itself so now we divide both sides by 2, and there's the answer!

$$c = £4{\cdot}40$$

£4·40 for a coffee? No wonder Malcolm looked shocked.

Dos and Don'ts

Algebra has a few more little rules that sound confusing, but they make more sense if you imagine we've got a set of identical matchboxes. Each box contains m matches so if we have a pile of three matchboxes, the total number of matches will be $3 \times m$ or just $3m$. The number 3 here is the *coefficient* of m.

Now we've got our matchboxes organized, we'll look at the rules in turn and then see how they apply to the matches.

❶ You can multiply a coefficient by a number.

If we bring on a second pile of three matchboxes . . .

You'll see that from 2 lots of $3m$ we get $6m$.

❷ You can't add a number to a coefficient.

If we find three spare matches . . .

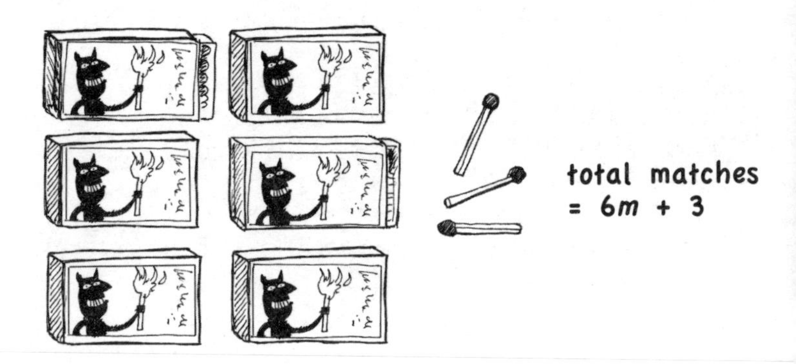

You'll notice that we have $6m + 3$ matches. You can't add the 3 to the 6 and get $9m$!

❸ You *can* add coefficients if the letter is the same.

If we find two more matchboxes . . .

You'll notice we can add the $6m$ to the $2m$ and get $8m$, but we can't add in the 3.

Here are three more rules, but don't worry if they don't make sense now. You'll see how they come in later on.

❹ If you have a minus outside a bracket, when you get rid of it you change the signs inside.

If you have something like $3 - (2x - 4)$, everything in the bracket has to be multiplied by -1. When you get rid of the bracket you get $3 - 2x + 4$. The $+2x$ has gone to $-2x$ and the -4 has gone to $+4$.

❺ If you multiply a letter with itself it becomes squared.

So if you had $y \times y$ it comes out at y^2. (We saw what squared numbers are back on page 64.) If you have $4y \times 3y$ you get $12y^2$. The coefficients just multiply and the letters get a squared sign.

❻ If you multiply different letters and numbers, you multiply the numbers, then just write the letters together afterwards.

So if you have $2x \times 4y$ it gives you $8xy$. This sort of thing tends to come up when you multiply things in brackets. For instance: $3p(7q - 2p) = 21pq - 6p^2$.

Right, let's see what this stuff can do for us.

Maths Mysteries Explained with Algebra

Algebra can come in incredibly handy for when you're solving problems or working out puzzles. Here are a few to get you started:

The Cornfield Con

Botchup Buildings wants to buy a plot of land from Farmer Sharpe. They agree it should be a square measuring 20m × 20m, which would have a total area of 400 square metres or m^2. However, when Botchup comes to inspect the plot, he finds something wrong: the field is rectangular instead of square!

Is the farmer being fair?

Even though we don't know how many metres too short or long the sides are, we do know it's the same number, so we'll call it x. We can draw a diagram of the field.

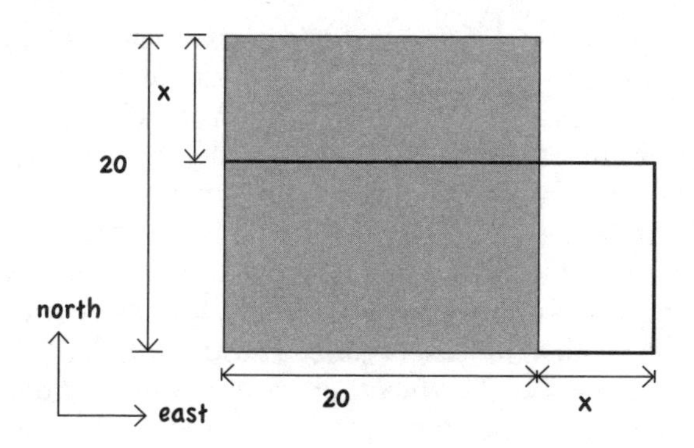

The grey area shows how the field would have looked as a 20m × 20m square. The rectangle measures $(20 - x)$ going north and $(20 + x)$ going east. To get the area of the rectangle we multiply these two numbers together and get:
$(20 - x) \times (20 + x)$, but we'd usually miss the × out and just put $(20 - x)(20 + x)$.

When you multiply two lots of brackets, you have to multiply everything in one bracket by everything in the other.

The way to do this is open the first bracket and multiply each bit by the second bracket. The sum looks like this:

$$
\begin{aligned}
(20 - x)(20 + x) &= 20(20 + x) - x(20 + x) \\
&= 400 + 20x - 20x - x^2 \\
&= 400 - x^2
\end{aligned}
$$

You'll see that when we multiplied out the $-x(20 + x)$ bit, we first had to work out $-x \times 20 = -20x$. Notice that we need to hang on to the minus sign. Finally we had $-x \times x$ which gave us $-x^2$. On the next line the $+20x$ and $-20x$ cancel each other out, leaving us with the curious answer of $400 - x^2$. What does this tell us?

Botchup should have had 400 square metres of land if the plot had been a square. But as soon as the farmer altered the shape as he did, the area of the plot was reduced by x^2. The bigger the value of x, then the more Botchup is being cheated. (Remember that x is how much longer/shorter the sides are.)

If the field was 5 metres too short one way and 5 metres too long the other, then $x = 5$. We can work out the area of the rectangular plot in two ways. First we can use our answer of $400 - x^2$ by changing the x for 5. The area will be $400 - 5^2$, which is $400 - 25 = 375$. The other way is to multiply the sides of the rectangle. Going north it's $20 - 5 = 15$ and going east it's $20 + 5 = 25$. The area is $15 \times 25 = 375$. Both these answers are the same so the algebra has worked!

The Difference of Two Squares

You have a square sheet of stamps measuring 6×6. Somebody decides to take a few and leaves you with a square measuring 4×4. How many have they taken?

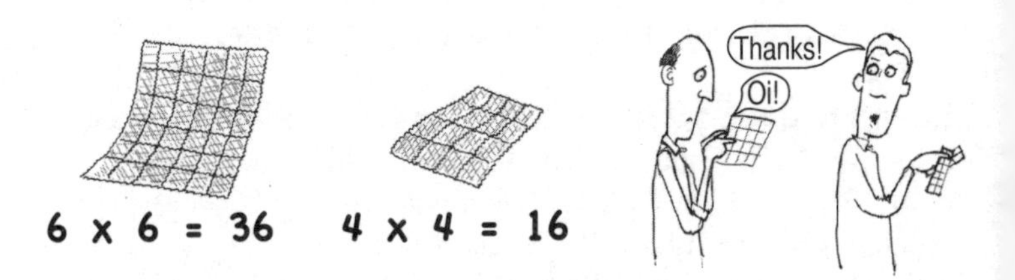

We need to work out $6^2 - 4^2$. If you subtract one square number from another it's called *the difference of two squares.* This one isn't too difficult because the numbers are small. We get $36 - 16 = 20$. However there's a short cut that works for any two squares.

The difference of two squares equals the sum times the difference.

The wording is very confusing but here's what it's saying. To work out $6^2 - 4^2$ you first need the *sum* of the two numbers: $6 + 4 = 10$. You also need the *difference* of the two numbers: $6 - 4 = 2$. You then multiply the sum by the difference. Sure enough $10 \times 2 = 20$, which is what we worked out before.

Instead of using words, it's easier to write the 'difference of two squares' rule out using an algebra equation. We'll use a to represent one of the numbers and b to represent the other, and here comes the rule:

$$a^2 - b^2 = (a + b)(a - b)$$

We've seen how this works when $a = 6$ and $b = 4$, but this equation will work for any values of a and b. If you're thinking that the difference of two squares would never turn up in real life, imagine $a = 20$ and $b = x$, then look back at the equations in the Cornfield Con. You'll see we got $(20 - x)(20 + x) = 400 - x^2$, so there it is!

The Three-Number Trick Explained

Back on the times tables grids on page 17 we saw that you can pick any three consecutive numbers, then if you multiply the smallest and biggest, the answer always comes to one less than the middle number squared. For instance, if we pick 12, 13 and 14 then $12 \times 14 = 168$, which is one less than $13^2 = 169$.

Once again, we'll use the equation that describes the difference of two squares, but we'll swap the *b* for the number 1. Here's what we get:

$$a^2 - 1^2 = (a + 1)(a - 1)$$

The first thing to notice is that $1^2 = 1 \times 1 = 1$, so we'll quickly put that in:

$$a^2 - 1 = (a + 1)(a - 1)$$

Now suppose *a* is the middle number of our three numbers. $(a + 1)$ will be the biggest number and $(a - 1)$ will be the smallest number. The equation is telling you that if you square the middle number to get a^2 and then take away 1, it equals the biggest number and the smallest number multiplied together.

When we did the trick with 12, 13 and 14 then $a = 13$, but of course *a* could equal any number you like, so that's why this trick works with any three consecutive numbers.

As soon as you have a mystery involving 'pick any number' then you can use algebra to investigate it.

How to Make the Universe Collapse

Do you remember back on page 74 I warned you this was possible? If you've got this far then you've worked hard and done well, so it's only fair that you should be rewarded with unlimited cosmic powers . . .

Start with any two numbers, *a* and *b*, which just happen to be the same:

$$a = b$$

We are going to treat both sides of this equation exactly the same all the way through. Watch carefully . . .

Multiply both sides by a: $a^2 = ab$
Subtract b^2 from both sides: $a^2 - b^2 = ab - b^2$

On the LHS we have the difference of two squares so we know that $a^2 - b^2 = (a + b)(a - b)$. On the right-hand side we have $ab - b^2$. Both bits will divide by b so we can write this as $b(a - b)$. This is all perfectly acceptable and correct.

So we get this: $(a + b)(a - b) = b(a - b)$

Now we divide both sides by (a – b) to get: $(a + b) = b$

There is nothing to multiply the brackets by, so we can get rid of them:

$$a + b = b$$

Move the + b across and change the sign: $a = b - b$

So we have: $a = 0$

Remember that a and b could be *any* number, so we've just proved that any number equals zero. This implies that any measurement of time, distance or weight is non-existent, so bye-bye universe!

The fault is when we divided both sides by (a – b). If a = b then (a – b) = 0. The one thing you cannot do with equations is divide both sides by zero! That's unless you fancy a day off work trying to make the universe collapse . . .

Simultaneous Equations

If you have two numbers that you don't know, you can usually work them both out if you have two different equations.

Here's a classic puzzle. A pair of shoes and a cleaning cloth cost £51. The shoes cost £50 more than the cloth. How much was the cloth?

Try asking Malcolm this; with luck he'll say that the cloth is £1 and the shoes are £50, but that would mean the shoes only cost £49 more than the cloth, so he's wrong!

If you think it through, you might get the answer, but here's how a bit of quick algebra works it out. We'll say the cost of the shoes is s and the cloth is c. Luckily this information is enough to give us the two equations we need:

Eq1. Shoes and cloth cost £51 so: $s + c = £51$

Eq2. Shoes cost £50 more than the cloth so: $s = £50 + c$

The most straightforward way of solving simultaneous equations is *substitution*. Eq2 tells us that $s = £50 + c$, so we'll write out eq1 again, but replace s with $£50 + c$.

Here's what we get: $£50 + c + c = £51$

We move the £50 over and change the sign, and also add the two c s: $2c = £51 - £50$

This bit's easy . . . $2c = £1$

And finally we divide both sides of the equation by 2 to get the cost of the cloth: $c = 50p$

Eq2 tells us that $s = £50 + c$ so we get: $s = £50{\cdot}50$

So the shoes cost £50·50 and the cloth costs 50p. It's a surprising answer, but it's the right one!

Think of a Number

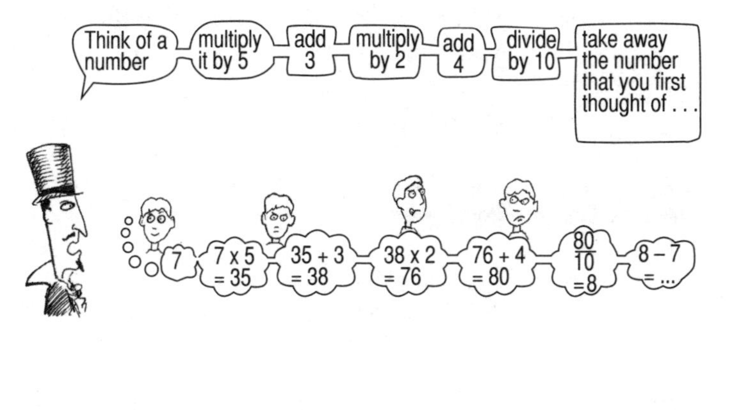

This trick will work with any number, or even a fraction. We can check this with algebra. As we have no idea what number will be picked, we'll just call it n and see what happens to it as we go through the instructions.

The Trick	What's Happening	Running Total
Think of a number	We call it n	n
Multiply by 5	So now we have $5n$	$5n$
Add 3	$5n + 3$. So far so good	$5n + 3$
Multiply by 2	We need to multiply everything so far by 2 so we put it in a bracket to make sure	$2(5n + 3) = 10n + 6$
Add 4	That's easy enough	$10n + 6 + 4 = 10n + 10$
Divide by 10	Everything has to be divided, so we'll use another bracket	$(10n + 10) \div 10 = n + 1$
Take away the number you first thought of	We just subtract n from the end	$n + 1 - n = 1$
Your answer is 1	n has completely disappeared and 1 is all that's left!	1

Time to Leave Algebra

If you open a school text book you'll see lots of x and y and problems which are all just a matter of shuffling things around and using a bit of common sense. There are massive books devoted to what you can do with algebra, so obviously we can't cover everything here, but here's one last problem that algebra solves rather neatly:

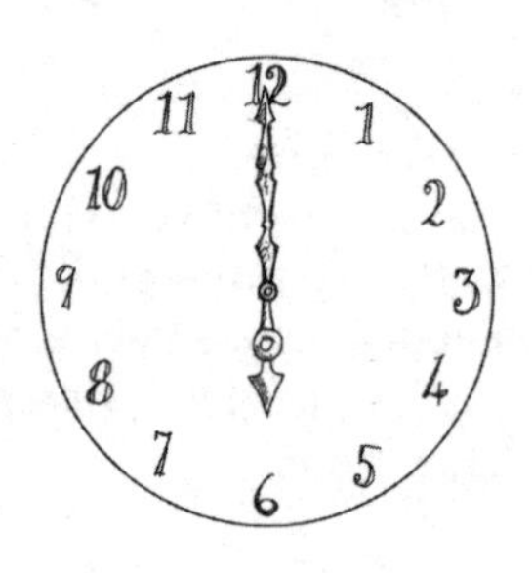

The time on the old clock is exactly 6 p.m. What time will it be when the minute hand next crosses over the hour hand?

Of course, the problem is that the hour hand is slowly moving all the time, so how can we allow for it?

Let's say that *m* is the number of minutes after 6 p.m. when the minute hand crosses the hour hand.

The minute hand has to move 30 minutes to get to the 6 position, and then it also has to move that extra bit of distance that the hour hand has moved in *m* minutes. Let's make this nice and clear:

distance moved by minute hand = 30 + distance moved by hour hand

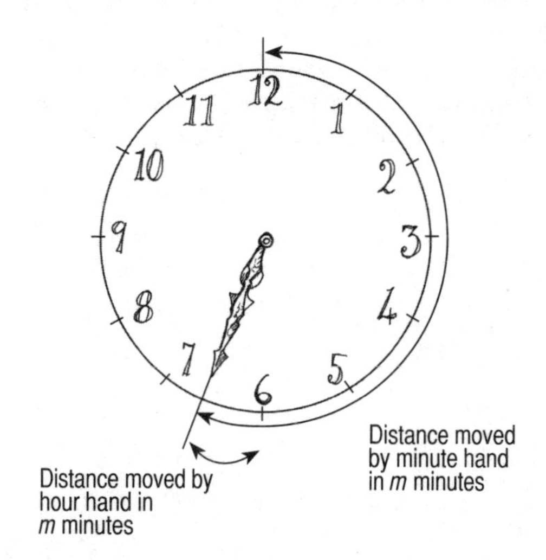

We need to know how far the hour hand moves in *m* minutes.

The minute hand takes 1 hour to go right round. The hour hand takes 12 hours to go right round, so it only goes $\frac{1}{12}$ of the speed. Therefore, when the minute hand has moved a distance of *m* minutes around the clock, the hour hand has moved $\frac{m}{12}$.

We can now make an equation: $m = 30 + \frac{m}{12}$

The 12 on the bottom looks nasty, but don't worry. We just multiply everything on both sides of the equation by 12: $12m = 360 + m$

Bring the $+m$ across and change signs: $12m - m = 360$

Take $1m$ from $12m$: $11m = 360$

Divide both sides by 11: $m = 32{\cdot}727$

So the answer is that the hands cross at 32·727 minutes after 6 p.m. However 0·727 of a minute is a bit ugly. As there are 60 seconds in a minute, the number of seconds is 0·727 × 60 which comes to about 44 seconds. Now we have a sensible answer: the hands will cross at 6:32:44.

SPEED

All of us need to plan a journey at some point. You might be wondering how long it should take you to drive to work, or maybe if you got to work too soon you wonder if you could have been caught by a speed camera . . .

Working Out Speed

Three factors are involved in journeys: distance, speed and the time it takes. Here is how they all link up:

distance = speed x time or $d = st$

It's easy to remember 'D Equals S T' because the letters DEST are in alphabetical order. This equation leads to two others:

If you divide both sides by t you get: $s = \frac{d}{t}$

If you divide both sides by s you get: $t = \frac{d}{s}$

Here's a situation when these formulas will come in handy:

The ferry sets sail in 3 hours and they have 150 miles to go. How fast should they be going to get to the port in time?

If you know two things, you can always work out the third, so let's see what speed the car needs to go at. We know $d = 150$ miles and $t = 3$ hours, so let's put these into $s = \frac{d}{t}$, which gives us $s = \frac{150}{3} = 50$mph. The speed is in miles per hour because we divided the number of miles by the number of hours.

You need to make sure you're using the same units for speed calculations!

Getting the Units Right

They've got ten minutes to get to the maternity ward and he's pedalling at a speed of 20mph. Are they going to make it? We need to know how long it will take to travel 3 miles at 20mph. We use the $t = \frac{d}{s}$ equation: $t = \frac{d}{s} = \frac{3}{20}$

As the speed is in miles per hour, this gives the time as $\frac{3}{20}$ hours, but we want to know the time in minutes. As there are 60 minutes in an hour, the time in minutes is $\frac{3}{20} \times 60 =$

9 minutes. They should arrive with just one minute to spare so let's hope there isn't a queue at the reception desk.

Combining Different Speeds

Suppose you have 8 hours to drive 400 miles. If you drive at the same speed all the way then you can use $s = \frac{d}{t}$ to tell you that your speed should be 400 ÷ 8 = 50mph.

Now suppose that you've just driven the first 200 miles at 40mph. How fast do you need to drive the last 200 miles if the journey is still to take 8 hours? The obvious answer might seem to be 60mph, but that would be wrong!

First you need to work out how much time you've got left. So far you've driven 200 miles at 40mph so use $t = \frac{d}{s}$ to find how long you've already been driving for: 200 ÷ 40 = 5 hours.

You only have 3 hours left to drive 200 miles, so your speed needs to be 200 ÷ 3 = 66·7mph.

PERCENTAGES

Percentages turn up everywhere from shops to banks, from wage slips to exam results. The simplest percentages, such as 50%, 33% or 25% are often used to describe special offers in shops, but you come up against more complicated percentages when you're working out taxes or using a credit card. It's well worth understanding how they work, and just like everything else in maths, it all starts off very simply. One per cent is written as 1% and it means exactly the same as or $\frac{1}{100}$ or 0·01. One hundred percent or 100% is $\frac{100}{100}$ which is the same as 1.

From Fractions and Decimals

If you want to convert a normal fraction to a percentage, you divide the bottom into the top and multiply by 100. Here's how you would work out $\frac{2}{5}$ as a percentage:

$$\frac{2}{5} = \frac{2 \times 100}{5} = \frac{200}{5} = 40\%$$

You can also convert back by dividing by 100. This is how you would find out 40% as a fraction:

$$40\% = \frac{40}{100} = \frac{4\not{0}}{10\not{0}} = \frac{\not{4}^{2}}{\not{10}_{5}} = \frac{2}{5}$$

Top and bottom divide by 10

Top and bottom divide by 2

Percentages and decimals are very closely linked because a percentage is the first two digits after the decimal point, e.g. 0·85 = 85%. If there is a zero after the point, then don't ignore it! 0·03 = 03% or just 3%. If the decimal has lots of digits then you just move the point along two places. Suppose you want to know what $\frac{1}{16}$ is, you put 1 ÷ 16 into your calculator and get 0·0625 which would be 6·25%.

Gradients

If you're cycling up a hill, the road sign tells you the steepness of the gradient using a percentage. A steeper road has a bigger percentage gradient. Here's how it's worked out:

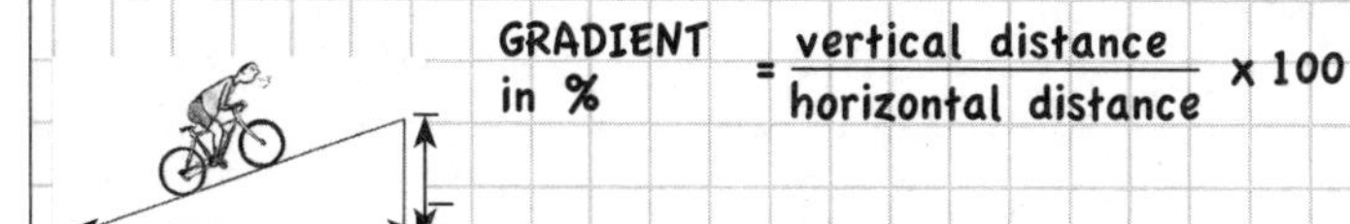

The vertical distance you go up is divided by the horizontal distance you go along. If you go up 1 metre for every 4 metres you go along, in the old days this would be called a gradient of 1 in 4. These days this gradient is made into a fraction of ¼ and then converted into a percentage. Therefore a gradient of 1 in 4 would be shown on a sign as 25%. It might not look much, but unless you're very fit, you'd need to get off and push.

When fractions are made into percentages most of them can't be converted exactly (just like converting fractions to decimals). Here are the most common percentages and the fractions they represent, with asterisks marking where the percentage has been rounded off:

$50\% = \frac{1}{2}$ $25\% = \frac{1}{4}$ $75\% = \frac{3}{4}$

$33\%^* = \frac{1}{3}$ $67\%^* = \frac{2}{3}$

$10\% = \frac{1}{10}$ $20\% = \frac{1}{5}$ $40\% = \frac{2}{5}$ $60\% = \frac{3}{5}$

$17\%^* = \frac{1}{6}$ $12{\cdot}5\% = \frac{1}{8}$

Money and Percentages

Most currencies link in very neatly with decimals and percentages. For instance, in Britain there are 100 pennies in 1 pound, in the Eurozone countries there are 100 cents in 1 euro and in the US there are 100 cents in 1 dollar. This helps to make the sums easy.

If you have £29 and split it between two people, each person gets £14$\frac{1}{2}$. Of course $\frac{1}{2}$ = 0·5 so you could say £14·5. This looks a bit odd, so we round pounds off to two decimal places. This gives each person £14·50 or £14 plus 50 pence.

As 100% = 1 and 100 pennies = £1, then 1% of £1 is just 1p. If a coffee and a cake in Barstucks cost £7 plus a service charge of 15%, it's not too hard to work out: 15% of £1 is just 15p. Therefore 15% of £7 is 15p × 7 = 105p. You'll end up paying £7 plus £1·05 = £8·05 and don't try fobbing them off with £8 because next time you go in there, you'll have no idea what they might sprinkle onto your whipped cream.

Spot the Bargain

Once you've got the hang of comparing fractions and percentages, you could save yourself some money. Let's suppose you set off to buy lots of batteries, and find that three shops have different special offers:

Batteries usually cost 50p each in each shop, so which is the best value? The trick is to find out what one battery costs in each shop.

30% Off: If they are knocking 30% off, then the cost is 70% of the normal price. 50p × 70% = 50 × 0·7 = 35p. (There's a little short cut to this sum. If you have 50 × 0·7, think of it as 5 × 10 × 0·7. You can then multiply 10 × 0·7 = 7. Your final sum is just 5 × 7 = 35.)

Buy 2 Get 1 Free: Here they are offering 3 batteries for the price of 2. The normal price of two is 2 × 50p = £1, but as they are giving you 3 batteries for this price, then each battery costs £1 ÷ 3, which is about 33p.

Buy 1 get a Second Half Price: The first battery costs 50p and the second costs $\frac{1}{2}$ × 50p, which is 25p. Therefore the cost for two is 75p. This means the cost for each battery is 75 ÷ 2 = 37·5p.

The cheapest battery was about 33p, so the best offer is 'Buy 2 Get 1 Free'. But then you go past another shop that usually sells batteries at the higher price of 65p, but they are having the well-known BOGOF deal – Buy One Get One Free. What does one battery cost?

Two batteries will cost 65p so one will cost $65 \div 2 = 32{\cdot}5$ p. This is the cheapest deal!

There's just one thing to be aware of: suppose it wasn't batteries you were buying, it was today's local paper. You only want one copy so there's no point buying two for 65p when you could just buy one for 35p at the 30% off shop. It's amazing how some people can't resist the best offer even when they don't actually want it!

Percentage Short Cuts

50%: divide by 2

25%: work out 50% then divide by 2

10%: divide by 10

5%: work out 10% and then divide by 2

$2\frac{1}{2}$%: work out 5% and divide by 2

1%: divide by 100

Once you've got used to these you can combine them to find lots of different percentages quickly.

15% of £25:	10% of £25 = £2·50 and so 5% = £1·25. Add these to get 15% = £3·75.
35% of £70:	50% of £70 = £35 so 25% = £17·50. 10% of £70 = £7 so 35% = £17·50 + £7 = £24·50.
$17\frac{1}{2}$% of £150:	10% of £150 = £15 and so 5% = £7·50 and $2\frac{1}{2}$% = £3·75. Add these three to get $17\frac{1}{2}$% = £26·25.

The Three Most Common Percentage Sums

Although it's convenient to do percentage sums with a calculator, it's sensible to have an idea what the answer should be rather than just trusting the screen. Don't get into arguments with waiters or shop assistants unless you're sure you've pushed the right buttons!

When you use the % button on a calculator you need to be especially careful. You don't usually need to push the = button afterwards. Calculator instructions are in < > brackets so you know when they start and stop:

❶ Of

What's 9% of £200?

Remember that *of* means multiply, so the sum is £200 × 0·09 = £18. You could think of the sum like this: 9% of £100 is £9. Therefore 9% of £200 is 2 × £9 = £18.

On a calculator you could try < 200 × 0·09 = > or < 200 × 9 % >

❷ Plus

What's £60 plus $12\frac{1}{2}$%?

This is the sort of thing you might find on a restaurant bill. The food/drink cost is £60, but they are adding on an extra service charge. To get the total, you work out $12\frac{1}{2}$% of £60 and then add it to the £60. You can work it out like this: 60 × 0·125 = £7·50 then £60 + £7·50 = £67·50.

Your calculator should be able to deal with this all in one operation if you push these buttons: < 60 + 12·5% >

There is a slightly cleverer way of doing this sum. The bill without the service charge is £60 × 100% which is just £60. The bill with the service charge is £60 × (100% + 12·5%). This comes to £60 ×112·5% or if you prefer £60 × 1·125 = £67·50.

❸ Less

What's £160 less 20%? If you are being given a special offer or discount you need this sum to find the final cost. The same sum would apply if you earned £160 but had to pay 20% in tax.

You could work out £160 × 20% and then subtract it from £160. The sums are £160 × 0·2 = £32 and then £160 – £32 = £128.

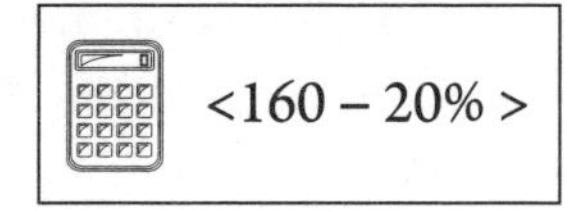

The other way of doing this sum is to think, instead of paying 100% of the price, you're only paying 100% – 20% = 80%. So the final cost is £160 × 80% = £128.

Percentage Mistakes

It's easy to fall into common traps with percentages. Here are two that you may come across from time to time:

Adding On and Taking Away

The shop owner obviously wanted his price to go back to £20, so what went wrong?

When you start with a price and then do more than one calculation on it, remember that *the original price is always 100%.* All percentage calculations should be based on this price.

When the assistant added 40%, the new price was 140% of the original price (£20 × 140% = £28).

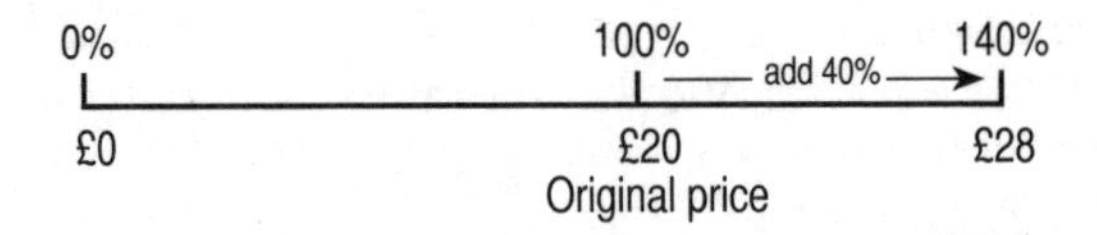

When she took off 40% she should have worked out 40% of the original price, and then taken that off. This would have brought the price back to 100% again. Her mistake was that she treated the new price as 100% and took 40% off that.

0% 60% subtract 40% 100%
£0 £16·80 £28
New price

Subtracting Added Tax

We've already seen how to add on a percentage charge such as service in a restaurant and we've also seen how to take a percentage off a sale price or tax off your wages (see the 'plus' and 'less' sections on pages 102 and 103). But suppose you just paid for a new laptop and the price, including the sales tax at 17·5%, came to £293·75. What would it have cost without the tax?

Nearly everybody makes the mistake of working out 17·5% of £293·75 and subtracting it. This would give you £242·34, which is wrong! It's the same mistake that the shoe-shop assistant made when she took 40% from £28.

The trick is to remember that the £293·75 is the final total you get when you start with the cost of the laptop without tax, and then multiply it by 1·175. Here's how you would write the sum:

$$\begin{pmatrix}\text{cost of laptop}\\ \text{without tax}\end{pmatrix} \times 1{\cdot}175 = \begin{pmatrix}\text{cost of laptop}\\ \text{with tax}\end{pmatrix}$$

If you're starting with the final total and working backwards to find the cost without tax, you *divide* the final total by 1·175:

$$\begin{pmatrix}\text{cost of laptop}\\ \text{without tax}\end{pmatrix} = \begin{pmatrix}\text{cost of laptop}\\ \text{with tax}\end{pmatrix} \div 1{\cdot}175$$

Now let's put the numbers in:

$$£293{\cdot}75 \div 1{\cdot}175 = £250$$

So the laptop was £250 before the tax was added. If you're able to claim the tax back, then you've got £43·75 coming to you.

INTEREST

Everybody knows that if you spend money you get poorer, but if you can manage to save money then you'll get richer… especially if you understand how interest works.

If you put some money in a savings account, this is called a *premium* (or a *principal*). It should earn some extra money for you called *interest* and this is worked out using a percentage called the *interest rate*. There are two main sorts of interest: simple and compound.

Simple Interest

Let's say you put £700 into an account that offers to pay a simple interest rate of 6% p.a. (p.a. stands for *per annum*). This means that every year they will work out what 6% of your savings are and add it on. So if you have £700 at 6% p.a., after the first year the interest you will earn is: £700 × 6% = £700 × 0·06 = £42. At the end of the year your account should have £742 in it.

If you manage to keep the money in for a few years, how much interest will it make you? You can use the simple interest formula:

simple interest $= p \times r \times t$

p = **the premium**
r = **the rate (as a decimal)**
t = **number of years**

So if you have £700 in the bank at 6% for 3 years, the interest you'd earn would be £700 × 0·06 × 3 = £126. (Remember to convert your 6% into 0·06!) This is added on to your £700, making your total savings into £826.

Compound Interest (Or How to Make More Money)

Suppose you put your £700 in for a year at 6% and then took it all out. As we've seen you'd have £742. Suppose you then put it all straight back in again for one more year. This time the interest would be 6% of £742 which is £742 × 0·06 = £44·52. This is slightly more than you got in the first year, and your total savings would now be £786·52. If you took this straight out and then put it straight back in for the third year your interest would be even more: £786·52 × 0·06 = £47·19.

After three years, your total savings would be £833·71. By taking your money in and out, you've made an extra £7·71. This is because you've been earning interest on your savings, plus extra interest on your interest! This is called compound interest.

The good news is that your bank doesn't want you tramping in and out once a year withdrawing everything and putting it back, so they calculate your compound interest for you. If they add on the interest once a year they use this formula:

$$\text{compound interest (once a year)} = [(1 + r)^t - 1] \times p$$

Let's check our £700 at 6% for 3 years. We'll replace r with 0·06, t with 3 and p with 700:

$$[(1 + 0{\cdot}06)^3 - 1] \times 700$$

It's vital to work these bits out in the right order (see page 37). We do the sums inside the most buried brackets first, then powers, then multiply/divide then add/subtract:

$[(1 + 0{\cdot}06)^3 - 1] \times 700$ The buried bracket is $(1 + 0{\cdot}06)$ which comes to $(1{\cdot}06)$.

$= [(1{\cdot}06)^3 - 1] \times 700$ We then see the bracket has a power of 3 so we need to work out $(1{\cdot}06)^3$ which means $1{\cdot}06 \times 1{\cdot}06 \times 1{\cdot}06$. Don't be proud – reach for the calculator!

Powers on a Calculator

If you have something like $(1{\cdot}06)^3$ you can just push <1·06 × 1·06 ×1·06 = > and get the answer 1·191016. If you have a posh calculator, it might have a powers button marked x^y or x^y or even y^x. You can work out $(1{\cdot}06)^3$ by pushing <1·06 x^y 3 = > and you should get the same answer, 1·191016.

$= [1{\cdot}191016 - 1] \times 700$ Now that we've worked out the power, we can get rid of the inside brackets. This leaves a simple sum in the big brackets $1{\cdot}191016 - 1 = 0{\cdot}191016$

$= 0{\cdot}191016 \times 700$ We're nearly there . . .

$= 133{\cdot}71$ This tells us our interest is £133·71.

Our original savings were £700, so after 3 years we have £700 + £133·71 = £833·71. That's the answer we got earlier, so the formula works!

The banks actually use even more complicated formulas which update your interest every month or even every day, and that gives you a tiny bit more money . . . isn't it nice of them to go to all that trouble? Unfortunately it works both ways, as you'll know if you have a credit card!

Loan Interest (Or How to Lose It!)

If you borrow money or pay for something on a credit card, *you* pay *them* interest! The more you borrow, the more interest you pay.

There are thousands of different deals and interest rates depending on what you earn, what you want to borrow, what you want the money for, how fast you can pay back, whether they like the shirt you're wearing and so on. Bob Hope summed it all up beautifully: 'A bank is place that will lend you money if you can prove you don't need it.'

For this section I'm just going to assume that you've borrowed £5,000 from Shark Loans at the ridiculous interest rate of 10% per month. To clear your debt you need to pay back both the interest and the original £5,000, which is usually called the *capital*.

Exploding Interest . . . And Repayments

You might think that the interest for the year will be 10% × 12 = 120%. Add this to the capital of 100%, and you might think that at the end of the year (if you paid nothing back) you'd owe 220%. Sadly not.

Loan interest is *always* compounded. After the first month you owe 110%, after the second month you owe 110% × 110%, after the third month you owe 110% × 110% × 110% and so on. Remember that 110% is the same as 1·1, so how much will you owe at the end of the year? It's $1{\cdot}1^{12} = 3{\cdot}14$ or 314%. The interest on interest really builds up!

If you don't repay any of that £5,000 you borrowed, after 12 months you'll owe: £5,000 × 3·14 = £15,700. Of course, the way to avoid this number getting so big is to make regular repayments.

The lowest sensible repayment (if Shark Loans will let you) is *interest only*. Every month they see what you owe, work out the interest and then that's what you pay. 10% of £5,000 = £500, but if you pay them £500 every month, this will not reduce your capital. The £5,000 will be owing for ever and you'll be paying interest for ever.

If you pay back a bit more than the interest (let's say £600), the £5,000 will start getting smaller. At first it only goes down very slowly, but it speeds up over time and eventually you'll have cleared the debt. The numbers here have been rounded off to the nearest pound:

Month	Amount Owing at Start of the Month	10% Interest	Add the Interest	You Pay Back	Amount Owing at End of the Month
January	£5,000	£500	£5,500	£600	£4,900
February	£4,900	£490	£5,390	£600	£4,790
March	£4,790	£479	£5,269	£600	£4,669
April	£4,669	£467	£5,136	£600	£4,536
May	£4,536	£454	£4,990	£600	£4,390

There are three things to realize here.

❶ The interest added on is getting lower every month. This means that you will be paying off the capital faster and faster as the months go by.

❷ In these five months your total payments were £3,000, which is £500 more than if you just paid the interest only. However, you have reduced your debt to £4,390, so you've paid back £610 of the capital. That extra £500 has been worth £610 to you!

❸ If we continued this table you'd find that you would need to make 18 monthly payments of £600 and one final payment of £484 to clear the debt. The total repayment would be: £600 × 18 + 484 = £11,284.

If you can squeeze an extra £100 back each month, then the debt disappears much faster. You will only need to make 13 monthly payments of £700 and one final payment of £105 to clear the debt. The total repayment would be £700 × 13 + 105 = £9,205. By paying an extra £100 per month you would have saved over £2,000!

The Debt Spiral

The one real danger to avoid is a missed payment. Not only does your interest leap up, but those nice friendly people from Shark Loans will probably charge you a penalty. Let's say you're paying back the minimum £500, you miss the first payment and the penalty is £200.

This isn't pretty . . .

Month	Amount Owing at Start of the Month	10% Interest	Add the Interest	You Pay Back	Amount Owing at End of the Month
January	£5,000	£500	£5,500	MISSED! Add penalty of £200	£5,700
February	£5,700	£570	£6,270	£500	£5,770
March	£5,770	£577	£6,347	£500	£5,847
April	£5,847	£585	£6,432	£500	£5,932
May	£5,932	£593	£6,525	£500	£6,025

After 5 months you owe £6,025, so that one missed payment of £500 has already amassed over £1,000 in extra debt, and this will keep growing indefinitely. Even if you never miss another payment, after 12 months you'll owe £6,997, after 2 years it's £11,268, after 3 years it's £24,671 and after 4 years it's £66,738.

If Shark Loans hadn't charged you that £200 penalty, after 4 years you would owe £49,099. But of course Shark Loans *will* charge the penalty because after 4 years, that £200 has earned them almost £17,000 in extra interest!

When you hear about people getting into ridiculous debts, it usually arises from a few missed payments, penalties, and interest rates even higher than the examples I've used here.

MEASURING SYSTEMS AND CONVERSIONS

Before the metric system was devised in France over 200 years ago, people used hundreds of different measurements for length, weight and volume. The metric system rather cleverly manages to measure almost anything with just three basic units, which all link nicely together.

Metres, Litres and Grams

A metre was originally defined as 1/10,000,000 of the distance from the equator to the North Pole on a line going through Paris. This makes the distance from the equator to the North Pole 10,000km and the length around the equator about 40,000km. (Actually the Earth isn't perfectly round, and the equator measures 40,075km.)

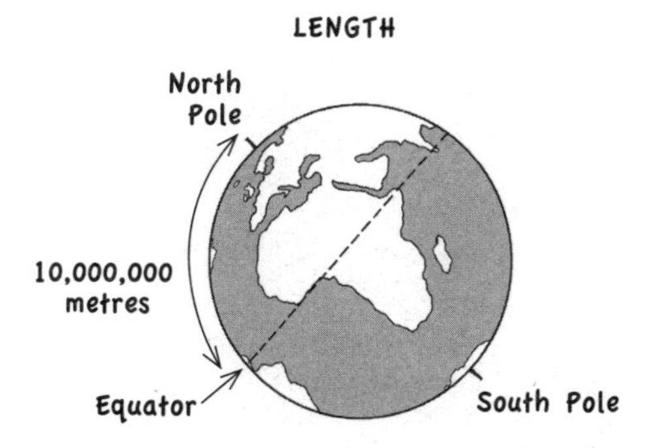

A gram is the weight of a cube of water measuring 10mm along each side at 4°C. It would be about the size of a small die. (The reason the water has to be at 4°C is that at this temperature water has its maximum density. In other words if the water was warmer or colder then the same sized cube of water would weigh a tiny bit less.)

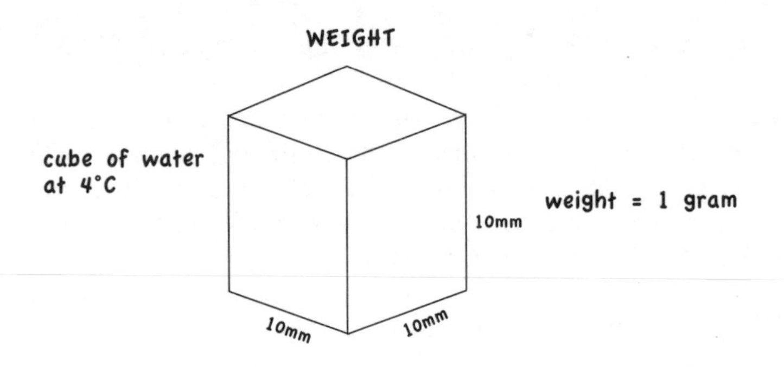

A litre is the amount of liquid you can fit into a cube measuring 100mm along each side. If your liquid is water at 4°C, then it will weigh exactly 1 kilogram. A cube measuring 1m along each side holds 1,000 litres of water, which weighs 1,000 kg, which is 1 tonne. In other words, 1 cubic metre of water weighs 1 tonne. If only everything in life was so simple . . .

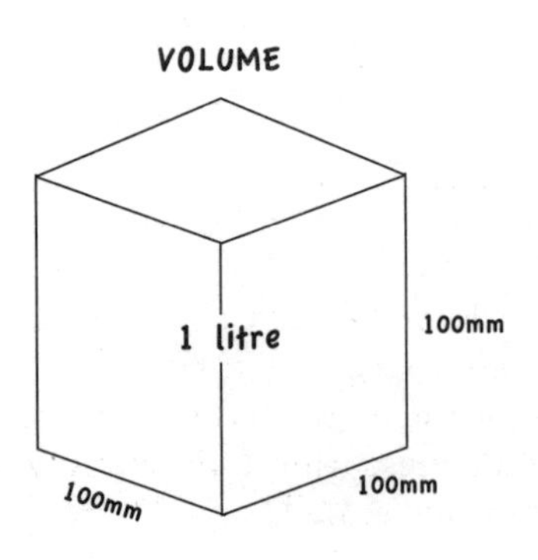

Although the weights are only dead accurate at 4°C, they don't vary too much. Even if the water is almost boiling, 1,000 litres will weigh about 0·96 tonnes, so that's pretty close.

Kilo, Mega and Milli

These basic units of metres, litres and grams can be multiplied or divided by thousands to give a different range of other measurements, from the very smallto the mind-bogglingly big. You'll probably know that 1 *kilo*metre is 1,000 metres and 1 metre is 1,000 *milli*metres, but here are how they fit in with the other prefixes:

Bigger Measurements	Smaller Measurements
deca (da) $\times 10$ or $\times 10^1$	deci (d) $\times 0{\cdot}1$ or $\times 10^{-1}$
hecto (h) $\times 100$ or $\times 10^2$	centi (c) $\times 0{\cdot}01$ or $\times 10^{-2}$
kilo (k) $\times 1{,}000$ or $\times 10^3$	milli (m) $\times 0{\cdot}001$ or $\times 10^{-3}$
mega (M) $\times 1{,}000{,}000$ or $\times 10^6$	micro (μ) $\times 0{\cdot}000001$ or $\times 10^{-6}$
giga (G) $\times 10^9$	nano (n) $\times 10^{-9}$
tera (T) $\times 10^{12}$	pico (p) $\times 10^{-12}$
peta (P) $\times 10^{15}$	femto (f) $\times 10^{-15}$

Remember that little minus sign in the power means 'divide' so $10^{-6} = \frac{1}{10^6}$. It's rare to use decametres, hectometres or decimetres these days, and even centimetres are seen as a bit outdated. A builder measuring a doorway won't usually say it's '75 centimetres' wide, he'll say '750 mill' which is short for millimetres. To convert millimetres to metres you just divide by 1,000 so 750mm = 0·75m.

Other Measurements You May Come Across

- Tonne: equal to 1,000kg or 1,000,000g or even 1Mg. A metric tonne is almost exactly equal to an old-fashioned British ton (but an American ton is lighter, about 0·9 of a metric tonne).
- Hectare: equal to an area of 10,000m^2, which is the same as a square measuring 100m × 100m. A hectare is about the same as 2·5 old-fashioned acres.
- Light year: you might not come across this in everyday life, but if you have an interest in astronomy, you'll know that a light year is the distance light can travel over one year, which is 9,500,000,000,000km. The nearest star to our sun is 4·2 light years away.

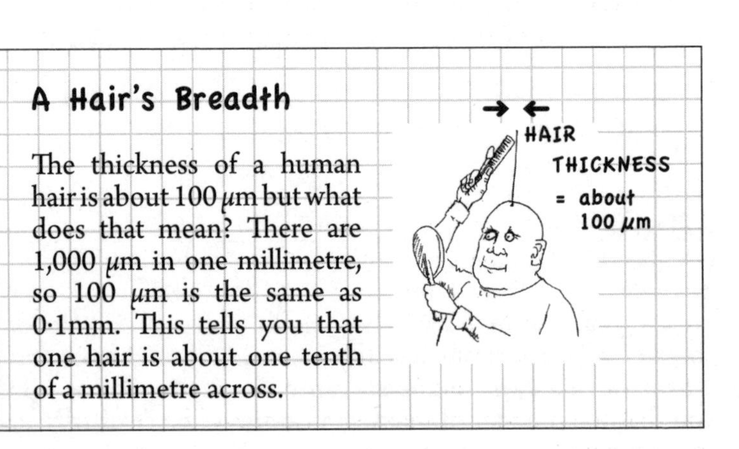

A Hair's Breadth

The thickness of a human hair is about 100 μm but what does that mean? There are 1,000 μm in one millimetre, so 100 μm is the same as 0·1mm. This tells you that one hair is about one tenth of a millimetre across.

Conversions

Converting between one unit and another happens all the time; whether you're cooking, buying clothes, changing currency, or looking at the weather forecast for next week. Here's a quick run-down of how to get the numbers right.

Converting from Imperial Units

Although the metric system is nearly everywhere, imperial units still crop up in odd places such as in old British or American cookery books or DIY manuals. Here's what you need to know if you need to convert between imperial and metric:

1 mile = 1·609 kilometres	1 British gallon = 4·55 litres*
1 yard = 0·914 metres	1 British pint = 568 millilitres*
1 inch = 25·4 millimetres	1 US gallon = 3.78 litres*
1 pound = 454 grams	1 US quart = 946 litres*
1 ounce = 28·35 grams	1 US pint = 473 millilitres*
1 fluid ounce = 28·4 millilitres	1 US cup = 237 millilitres*
1 acre = 4,047 square metres or about 0·4 hectares	(* These numbers only apply to liquid measurements.)

Note: 4 US cups = 2 US pints = 1 US quart = 0·25 US gallons.

For most conversions, you only ever need to multiply or divide; the clever bit is knowing which to do. Say you want to convert 28 kilometres into miles. If we don't need to be too accurate, we can say 1 mile = 1·6km, but do we work out 28 × 1·6 or is it 28 ÷ 1·6? You have to decide if the answer should be bigger or smaller. If you compare the 1 mile and 1·6km, obviously the number 1 is smaller than 1·6. This tells you that the number of miles should always be smaller than the number of kilometres.

As 28 ÷ 1·6 gives the smaller answer, this is the right one so 28km = 17·5 miles.

Metres and Millimetres

This should be *so* simple but it's so easy to get wrong! 1m = 1,000mm, so you either need to multiply or divide by 1000.

What's 0·04 metres in millimetres? Obviously you'll be expecting more millimetres so you just multiply: 0·04 × 1,000 = 40mm.

What's 520mm in metres? You're expecting fewer metres so you divide: 520 ÷ 1,000 = 0·52 metres.

Currency

Exchange rates work just the same as any other conversions. As I've no idea where you might be, or what you're changing into what, I'll assume you're a seventeenth-century pirate and you want to change 500 doubloons into groats.

If it's 1db = 3·76gr then you should expect a bigger number of groats. You should get 500 × 3·76 = 1,880 doubloons.

The only thing to watch out for is that a bank may also charge a commission for doing it, so you might not get as much as you expect. In that case do us all a favour and make them walk the plank. Yarrrr, me hearties!

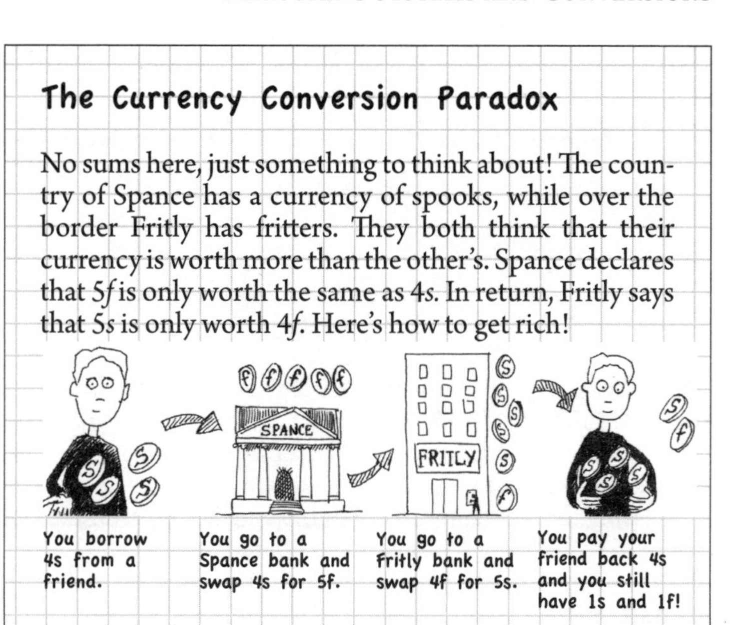

The Currency Conversion Paradox

No sums here, just something to think about! The country of Spance has a currency of spooks, while over the border Fritly has fritters. They both think that their currency is worth more than the other's. Spance declares that 5*f* is only worth the same as 4*s*. In return, Fritly says that 5*s* is only worth 4*f*. Here's how to get rich!

So you've made a profit, but who was the loser?

Temperature

The only common conversion that's a bit more complicated is temperature. Most of us now use Celsius (or Centigrade), but there's also Fahrenheit, and scientists use Kelvins.

	Fahrenheit	Celsius	Kelvin
Boiling point of water	212°F	100°C	373°K
Freezing point of water	32°F	0°C	273°K
Blood temperature	98°F	37°C	310°K
Absolute zero	–459·67°F	–273·15°C	0°K

Here's how to convert between °C and °F:

$°C = (°F - 32) \times \frac{5}{9}$	If you start with Fahrenheit, subtract 32, then times by 5 and divide by 9 and end up with Celsius.
$°F = (°C \times \frac{9}{5}) + 32$	If you start with Celsius, times by 9, divide by 5 and then add 32 to get Fahrenheit.

Absolute zero is the lowest possible temperature, which in Celcius is minus 273·15 °C. Kelvins are the same as Celsius, but they start at absolute zero. Therefore °K = °C + 273 (we can ignore the extra 0·15°C).

Getting the little sums in the right order is confusing, so the easiest way to check if you've remembered how to get °F from °C is to start with 100°C and see if you can convert it to 212°F. A bit of practice tells you that you multiply by 9 (to get 900) then divide by 5 (to get 180) and then add 32 (to get 212.)

Strange but true: −40°F = −40°C.

LINE, AREA AND VOLUME

If you bake a cake, tie a ribbon round it, then wrap it in cling film, you've had to deal with line (the length of the ribbon), area (the amount of cling film you need) and volume (the amount of ingredients in the cake). Suppose you divide a big bowl of cake mix into three equal lots, and then use them to make three different shaped cakes.

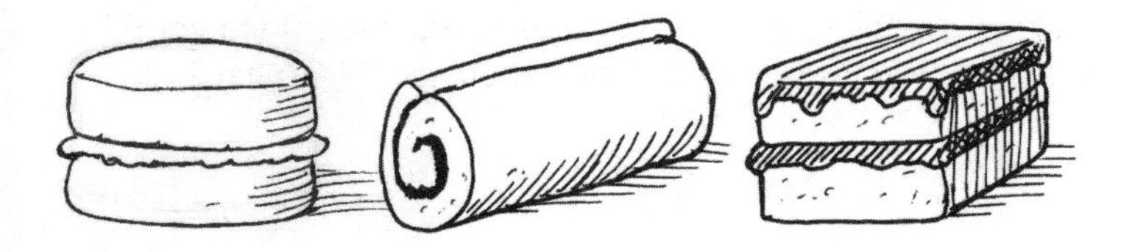

Although the three volumes will be the same, the amount of cling film and ribbon you need for each one may well be different. This is because line, area and volume all have their own special jobs to do, so it's as well to know how they work, whether you're baking cakes or planning home improvements.

Line

A line is the distance between two points. It doesn't matter how long the line is, it only needs one measurement. A pencil might be 130mm long, a running track might be 100m and a drive from Kilmarnock to Norwich is about 705km (or 438 miles). Millimetres (mm), metres (m) and kilometres (km) are all length measurements.

Let's say you want to fix a new TV socket and you need to buy some cable. You've taken a diagram of your living room into the shop.

You want to run a cable from an old socket round to your new socket. You have measured some of the walls, but not walls *a* and *b*. How much cable do you need?

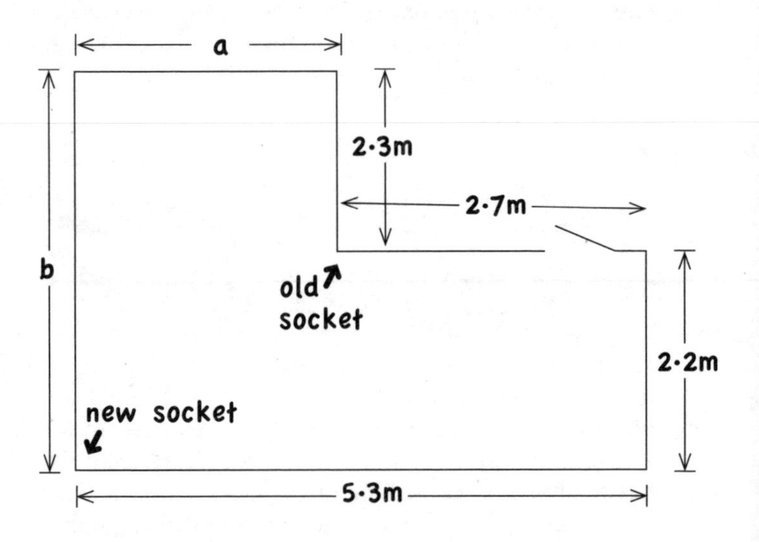

The coward's way out is to avoid the sums and run it past the doorway. The cable length would be 2·7 + 2·2 + 5·3 = 10·2m, plus you might need some more to go around the door. The clever way is to work out *a* and *b*. If you just look at the horizontal lines on the diagram, you can see that the long wall is 5·3m and the short wall is 2·7m. Therefore the wall marked *a* = 5·3 – 2·7 = 2·6m. For the vertical walls you'll find that *b* = 2·2 + 2·3 = 4·5m. If you run the cable round to avoid the doorway, the total length will be $2{\cdot}3 + a + b$. We then just put the numbers in for a and b to get $2{\cdot}3 + 2{\cdot}6 + 4{\cdot}5 = 9{\cdot}4$m.

Curved Space

A straight line is the shortest distance between two points . . . unless you're dealing with curved space! We usually think of a straight line drawn on a two-dimensional surface such as a flat piece of paper. If you've got a map of the world, you could draw a straight line to show the shortest flight path from the North Pole to the South Pole, but is that really the shortest distance between the two?

World Map in 2D

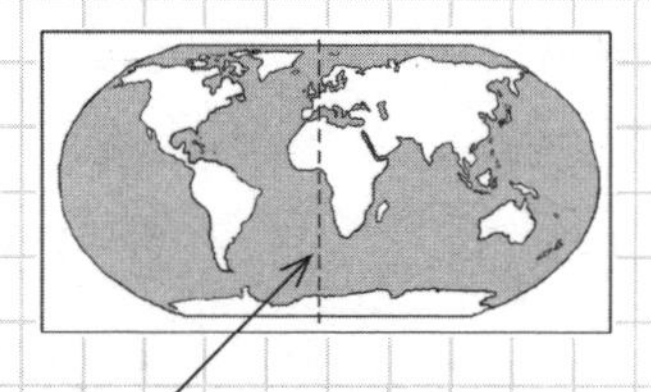

Shortest line between the two poles

World Map in 3D

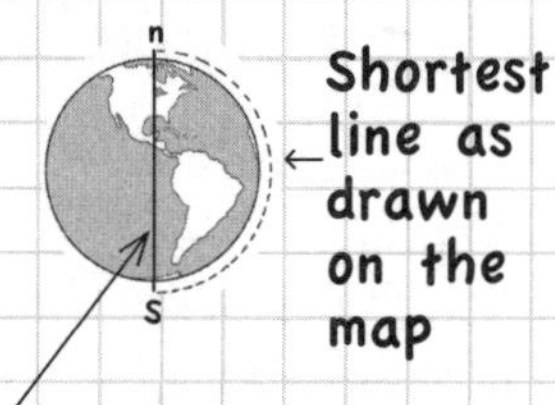

The real shortest line goes through the centre

When we look at the real world, we move up into three dimensions. Our flight path becomes a curve and the shortest line goes through the centre of the Earth. (We'll work out how long this line is later on.) If we knew how to move up to four dimensions, maybe we could find an even shorter way between the poles! This is the sort of thing Einstein was thinking about when he was studying his theory of relativity.

Area

Now that the line measurements are sorted, how much paint do you need to cover wall *b*, which we now know to be 4·5m long?

In the hardware shop you pick up a lovely tin containing 750ml of Burnt Eggshell. On the side it says that a litre of paint covers $12m^2$.

The first bit to notice is the m^2, which stands for *square metres.* This is a measurement of area. You can paint any shape you like, but in total 1 litre will cover the same space as 12 squares measuring 1m × 1m. So how much area will one tin cover? Well, 750 millilitres is 0·75 of a litre, therefore one tin will cover $12 \times 0{\cdot}75 = 9m^2$.

Finding the Area of a Wall

By now you'll have realized that knowing the length of the wall isn't enough. The amount of paint you need also depends on how high the wall is. There are a whole range of formulas to work out different shaped areas, and they all involve multiplying two lengths together in some way. We'll see some other formulas soon, but right now we're dealing with a rectangle, which is the most common shape, and luckily it has the easiest formula.

area of rectangle = width × height

Let's suppose our room is 2·5m high and we know the wall is 4·5m long. We just need to multiply the height by the length: $2{\cdot}5m \times 4{\cdot}5m = 11{\cdot}25m^2$.

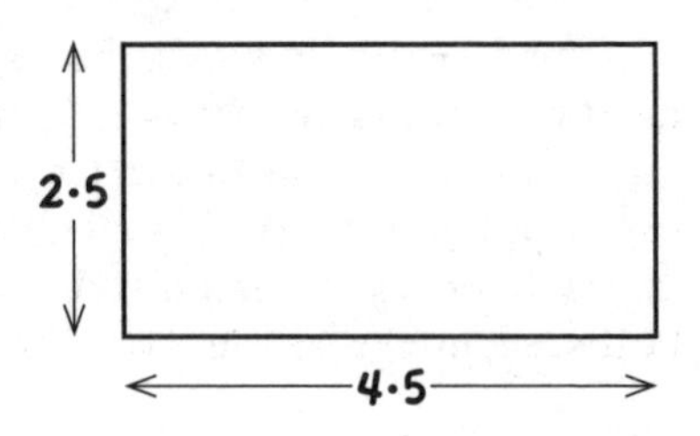

You'll see that when we multiply metres by metres we get square metres.

Now we're getting somewhere because we know the wall area is $11{\cdot}25\text{m}^2$ and one tin covers 9m^2. Without doing any more sums, we can see that one tin won't be quite enough to paint the wall. If we buy two tins there'll be a lot left over so let's go mad and see how much we need to paint all the walls. There are two ways to do this. One way is to work out the area of each wall in turn and then add them all up. As they are all 2·5m high we'd get this: $(4{\cdot}5\times2{\cdot}5)+(5{\cdot}3\times2{\cdot}5)+(2{\cdot}2\times2{\cdot}5)+(2{\cdot}7\times2{\cdot}5)+(2{\cdot}3\times2{\cdot}5)+(2{\cdot}6\times2{\cdot}5)$...yuck!

It's much easier to imagine all the walls set out in a straight line.

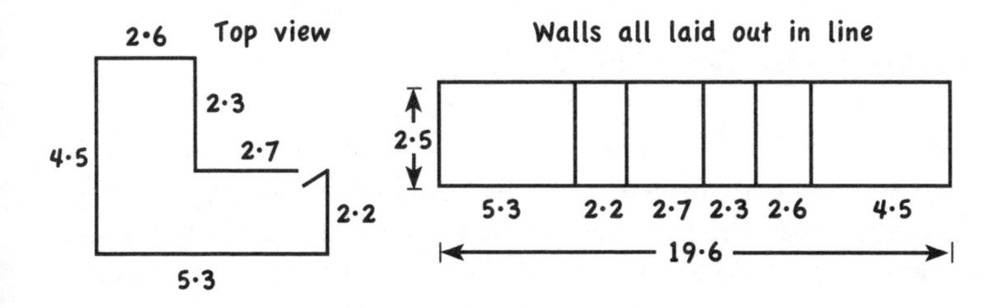

If you add up all the lengths of the walls you find that the total length around the room is 19·6m. You multiply this by 2·5 to get the total wall area: $19{\cdot}6\times2{\cdot}5=49\text{m}^2$.

Each tin will paint 9m^2 so the number of tins you need is $49\div9=5{\cdot}44$, so six tins will do it.

If you want to be even more accurate with the area you have to paint, you can measure up the doorway and windows and subtract them. As a short cut, a doorway usually measures $0{\cdot}75\text{m}\times2\text{m}=1{\cdot}5\text{m}^2$. Now imagine your windows. Are they as big as a door? Half a door? Two doors? Or, if you can't be bothered, just paint them over and save on curtains.

Bricks and Blocks

If you're using standard building bricks and you want to build a wall just one brick thick you need about 60 bricks per m^2. For blocks you need about 10 per m^2.

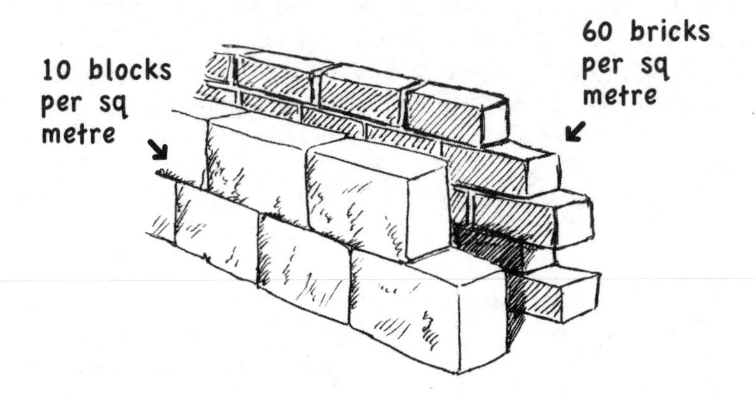

So if a runaway road roller happens to demolish your 5·3m by 2·5m wall, the number of bricks you will need for the outer wall is: 5·3 × 2·5 × 60 = about 800. And the number of blocks for the inner wall will be 5·3 × 2·5 × 10 = about 135. If my mate Blakey had known this it would have saved him a lot of embarrassment when the brickie he'd hired for the day ran out of materials by 11 a.m.

Painting the Ceiling

This Burnt Eggshell colour is so very lovely that you decide to paint the ceiling with it too, so we need to work out another area. Sadly it's not a perfect rectangle, but we can split it into two rectangles, work them out separately and then add them up. There are two ways of doing this:

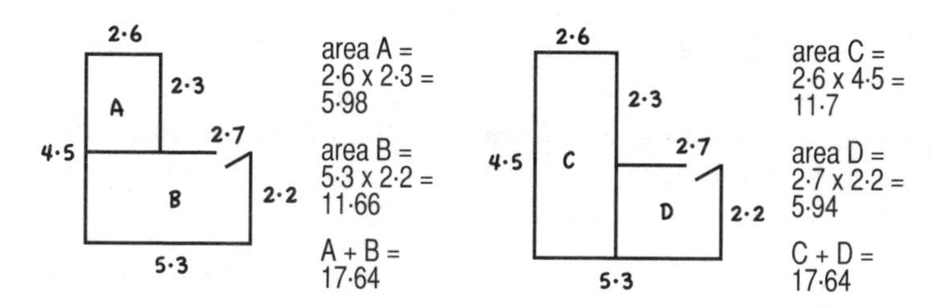

Now we know the area of the ceiling is 17·64 m^2. As each tin of paint covers 9 m^2, two more tins should just about do it.

Other Area Formulas

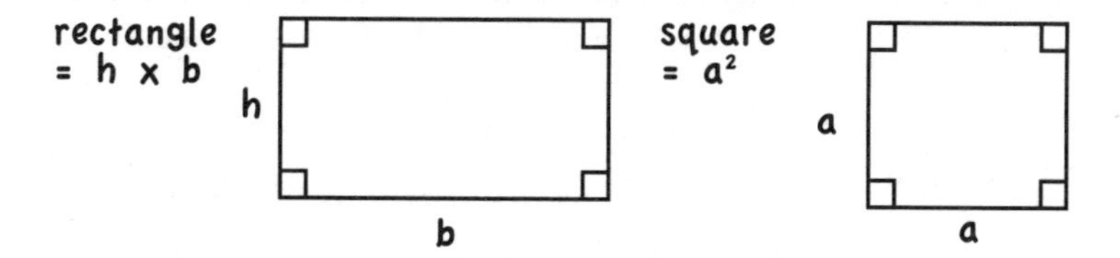

We've already used the rectangle formula. A square is just a rectangle with the sides the same length, so to get the area you just multiply one of the sides by itself.

A right-angled triangle is just a rectangle cut in half diagonally. Rather conveniently the area = $\frac{1}{2}$ × the two shorter sides of the triangle multiplied together.

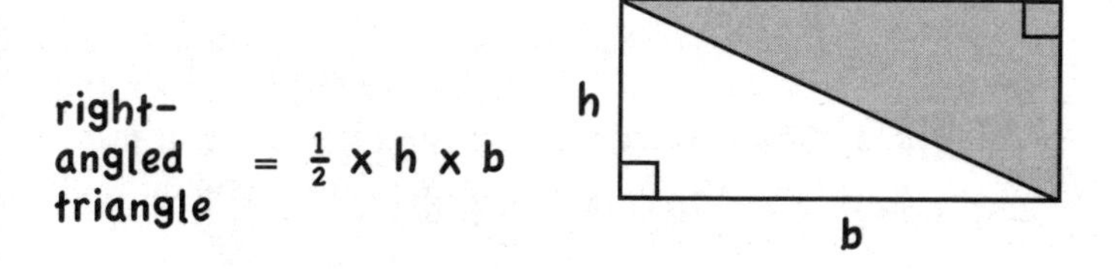

In fact *any* triangle is half the area of the smallest rectangle it fits into. Here you'll see the two grey areas will fit over the two parts of the triangle. This formula is described as: area = $\frac{1}{2}$ × base × perpendicular height.

any triangle = $\frac{1}{2} \times h \times b$

Even though triangle formulas are very popular in maths lessons it's unlikely you'll ever have to use them in the real world. Here are two more formulas that are even more useless unless you have some very odd shaped walls and ceilings.

trapezium = $\frac{1}{2}(a + b) \times h$

parallelogram = $h \times b$

The little arrow heads indicate parallel lines. For both of these formulas you need to know the perpendicular height or you're stuck.

Finally, here's how to find the size of an area on a map. If the squares represent 0·1km along each side, then the area of each is $(0{\cdot}1)^2 = 0{\cdot}01\text{km}^2$. Suppose a lake covers about 35 squares, the area is about $0{\cdot}35\text{km}^2$.

Any old shape you like – just count the squares!

Volume of a Cuboid

A line needs one measurement, an area needs two measurements multiplied together, and a volume needs three. The simplest volumes to work out are of *cuboids,* which are boxes with rectangular sides. You just multiply the length, width and height to get the volume.

If you're converting your spare bedroom into a giant aquarium for your pet octopus, you might want to know how much it will weigh when full of water. First you work out the volume, so if the room measures 4m × 3m × 2·5m, the volume would be 4 × 3 × 2·5 = 30m^3. You'll see volume is in m^3 or *cubic metres*. This volume of 30m^3 is the same amount of water as you'd get in 30 cubes each measuring 1m along each side. As we know from page 114 that one cubic metre of water weighs 1 tonne, your spare room will weigh 30 tonnes. Good luck!

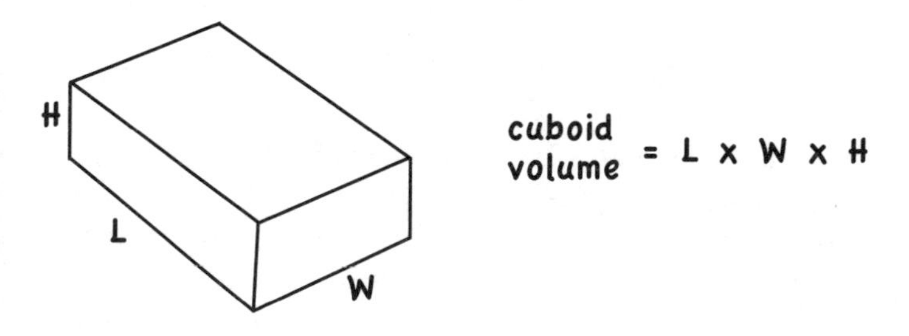

The only other volumes you're ever likely to need are cylinders, but for this we first need to meet π.

Circles and π

The outside of a circle is called the *circumference,* the distance across the centre is the diameter, the distance from the centre to the edge is the *radius* and they are linked up by π. This is the Greek letter called *pi* and it represents the special number you get if you divide the circumference of any circle by the diameter.

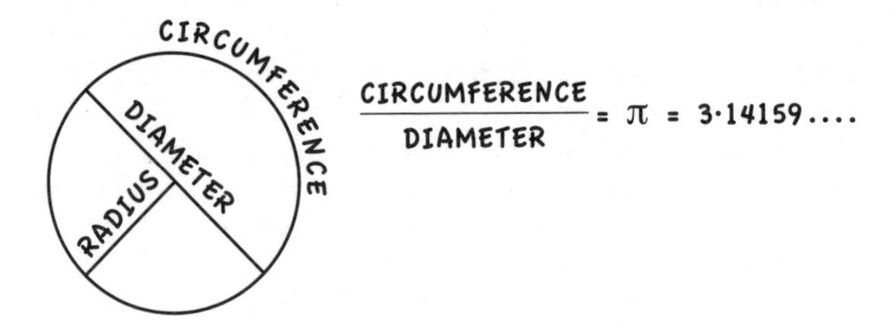

The decimal goes on for ever, but for most of us it's enough to remember it as 3·14 or as the fraction $\frac{22}{7}$. You need π to calculate the area of a circle or the volume of a cylinder or a sphere.

Working Out π

π has fascinated people ever since maths began because it is so hard to calculate exactly. The ancient Greek Archimedes drew a shape with ninety-six tiny straight sides that was almost a perfect circle, and from that he worked out that π was somewhere between $3\frac{10}{71}$ and $3\frac{1}{7}$, so he had it to within 0·001 of the exact answer.

In the sixteenth century the German Ludolph van Ceulen used a shape with over 32 *billion* sides and spent twenty years finding the first thirty-five digits of π. He probably thought that nobody would ever do better than this but, sadly for him, just after he died, people like Isaac Newton found easier ways of getting even more digits. These days, computers have found trillions of decimals of π, and they've still hardly scratched the surface…

The main formulas for circles are:

> **diameter = 2 x radius (usually written as $d = 2r$)**
> **circumference = π x diameter ($c = \pi d$ or $c = 2\pi r$)**
> **area = πr^2**

Do you remember the bit about straight lines on page 123? If you fly from the Earth's North Pole round to the South Pole, the distance is 20,000km. How far would it be if you drilled a straight line through the middle of the Earth?

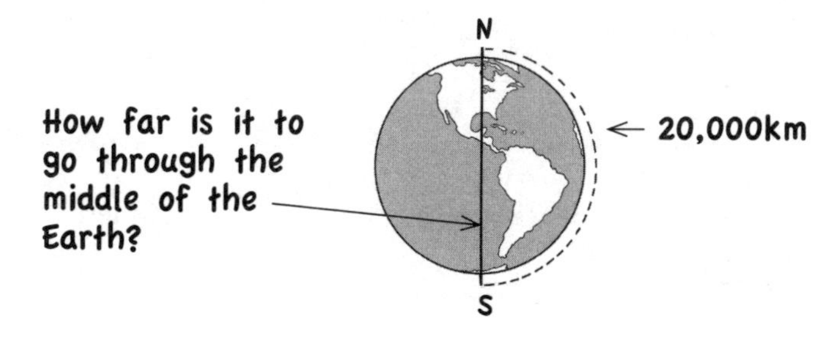

We know that 20,000km is half way round the Earth, so the complete circumference is $2 \times 20{,}000 = 40{,}000$km. The distance through the centre is the diameter d so we use the formula $c = \pi d$. We know $\pi \times d = 40{,}000$. Therefore $d = 40{,}000 \div \pi = 12{,}732$km.

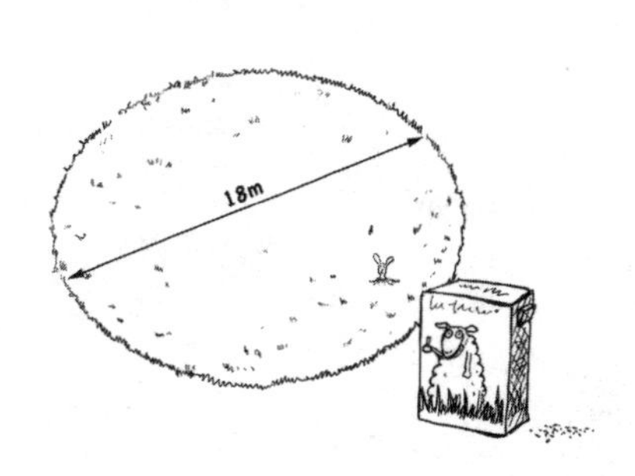

Let's try something closer to home: you're buying some grass seed for a circular lawn that measures 18m across. One box of seed covers $10m^2$. How many boxes do you need? We need to work out the area of the lawn so we use the formula *area* = πr^2. We know the lawn is 18 metres across, but this is the diameter. For the area formula we need the radius, which is half the diameter. Therefore $r = \frac{18}{2} = 9$ metres. Now we can work out the area using $\pi r^2 = \pi \times r \times r = \pi \times 9 \times 9 = \pi \times 81 = 254\ m^2$. As each box covers $10m^2$, you'll need $254 \div 10 = 25{\cdot}4$ boxes.

Cylinders

The main volume formula involving π is for a cylinder. You multiply the area of one end (which is a circle) by the height.

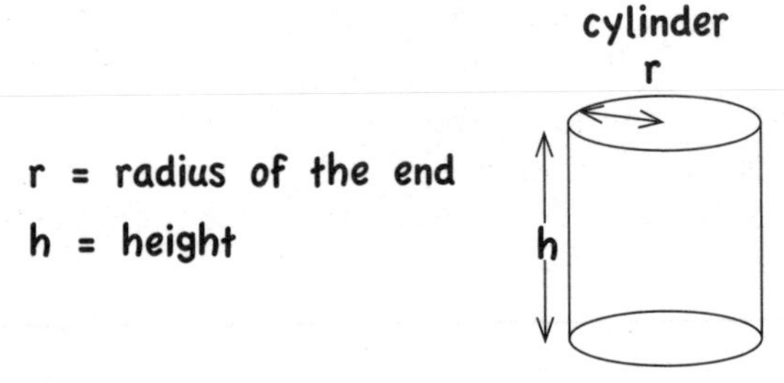

volume of a cylinder = $\pi r^2 h$

If you're in the hardware shop and they happen to have a huge economy tin of that Burnt Eggshell paint you love so much, it'd be handy to know how much it holds. Measuring the radius is tricky, so you measure across the top to find the diameter is 160mm, and then halve it to get the radius: $r = 160 \div 2 = 80$mm. You also measure the height to find it's 0·3m.

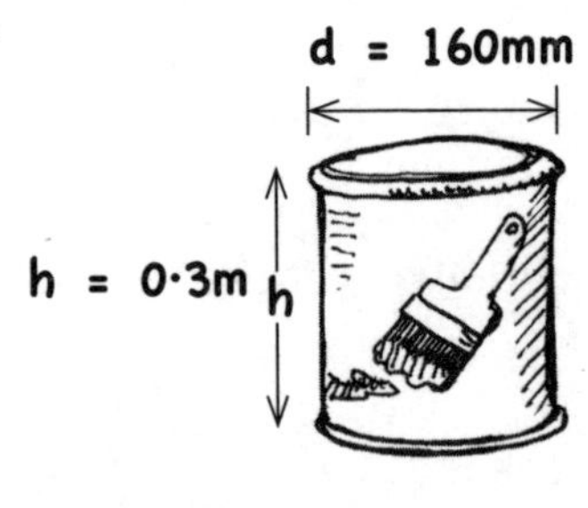

When you're working out area or volume, always use the same units for all the measurements. We've got the radius in millimetres and the height in metres, so let's stick with metres, which will give us an answer in cubic metres. We'll convert our radius of 80 millimetres into 0·08 metres. We now put $r = 0{\cdot}08$ and $h = 0{\cdot}3$ into the formula:

$$\begin{aligned}\text{volume of paint tin} &= \pi \times (0{\cdot}08)^2 \times 0{\cdot}3 \\ &= \pi \times 0{\cdot}0064 \times 0{\cdot}3 \\ &= 0{\cdot}00603\ \text{m}^3\end{aligned}$$

We now quickly turn back to page 114 and see that there are 1000 litres in 1 cubic metre, so this mystery tin holds $0{\cdot}00603 \times 1{,}000 = 6{\cdot}03$ litres.

Earlier we worked out that we'd need 6 of the 750ml tins for the walls and 2 for the ceiling, so that's 8 tins in total. How many litres is that? 750 millilitres = 0·75 litres so the 8 tins hold $8 \times 0{\cdot}75 = 6$ litres. This mystery tin is about the right size to do your walls *and* ceiling!

A Short Cut

If you have a measuring tape, you can work out the volume of a cylinder without worrying about π. You need to measure the circumference, diameter and height. The circumference automatically puts the π bit into the calculation and it makes for a lovely easy formula:

$$\text{cylinder volume} = \frac{dch}{4}$$

Spheres

About 2,250 years ago, the Greek mathematician and scientist Archimedes came up with the most extraordinary range of inventions. However, his favourite discovery is shown by the engraving on his tombstone: he was the first person to prove that a sphere occupies exactly $\frac{2}{3}$ of the smallest cylinder it will fit inside. In other words if you have a small tin of beans just big enough to fit a tennis ball inside, it will push out exactly $\frac{2}{3}$ of the beans. Thanks to him, we have a formula for the volume of a sphere.

You start with a sphere, and you call the radius *r*.

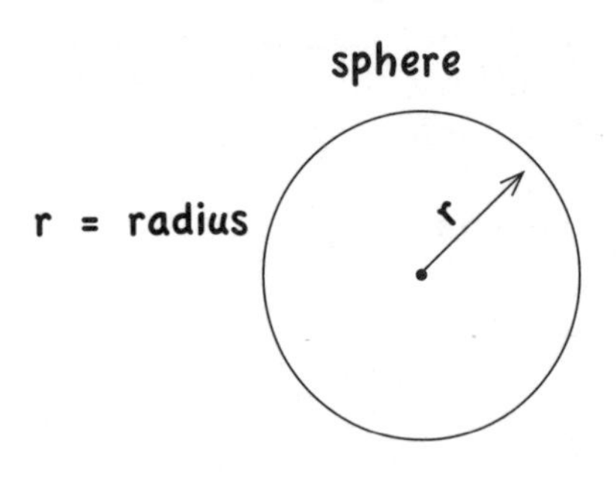

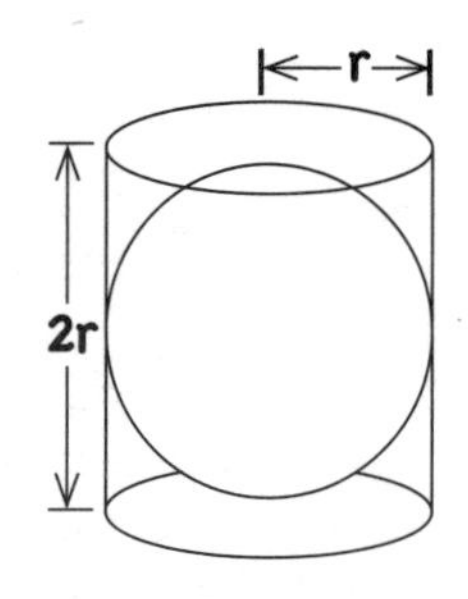

First we work out a formula for the volume of the smallest cylinder that the sphere can fit inside. We use the normal cylinder volume formula $\pi r^2 h$, but we can see that the height of the cylinder $h = 2r$. Therefore the volume of the smallest cylinder = $\pi r^2 \times 2r = 2\pi r^3$.

Archimedes showed that the sphere has $\frac{2}{3}$ of this volume, so the volume of the sphere = $\frac{2}{3} \times 2\pi r^3$. We end up with:

volume of a sphere = $\frac{4}{3}\pi r^3$

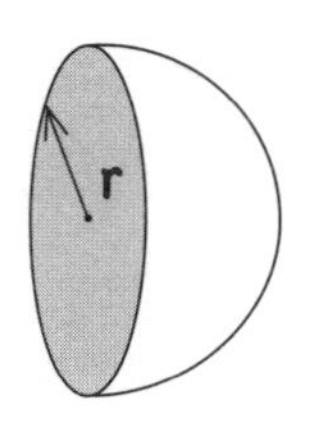

While we're at it, if you chop a sphere in half, the area of the flat circle is πr^2. The area of the whole outside of the sphere happens to be four times as big as the circle, so we get:

surface area of a sphere = $4\pi r^2$

The sphere volume is another formula that's very popular in maths lessons, but in the real world it's almost completely impractical: how are you supposed to measure the radius of something like a football from its exact centre? It's much easier to measure the circumference and then use this formula:

volume of a sphere = $\frac{c^3}{60}$

If you're a rocket scientist and need a more accurate answer then work out: $\frac{c^3}{59.2176264}$

But if you're a rocket scientist and you need to read this book then we're all in big trouble, aren't we?

Pythagoras and his Theorem

Pythagoras lived about 300 years before Archimedes, and this rule is what he is best remembered for: in a right-angled triangle, the square on the hypotenuse is equal to the sum of the squares on the other two sides.

This sounds confusing, but a diagram makes it clearer. If you have a right-angled triangle and draw a square on each side, then the areas of the two small squares will add up to the same as the big square.

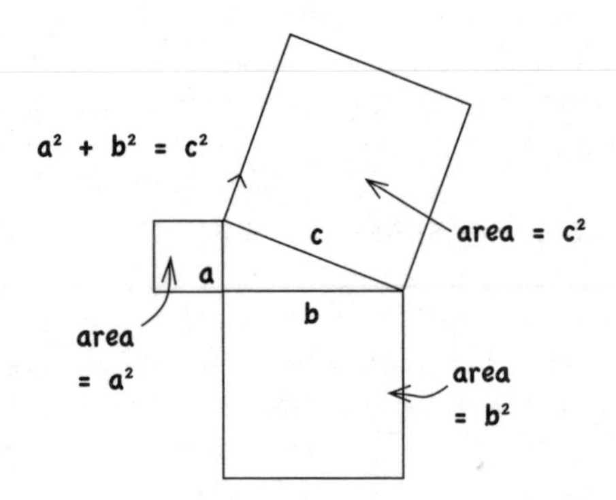

If you're wondering why anyone would ever need to stick squares on the sides of a triangle, don't worry. That's not why this theorem is useful. Instead, imagine you're taking a short cut diagonally across a football pitch. If the pitch measures 100m × 70m, how far do you have to walk?

There are a few fiddly sums coming up, so make sure the answer looks right! The diagram shows that the answer should be more than 100m but less than 170m.

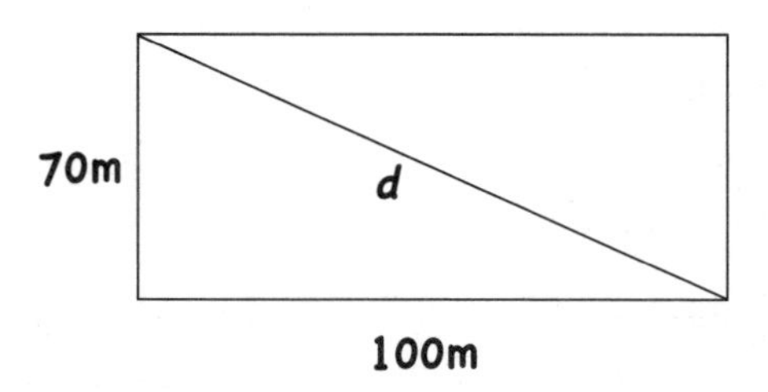

Let's call our diagonal short cut *d*.

According to Pythagoras: $d^2 = 100^2 + 70^2$
This becomes: $d^2 = 10{,}000 + 4{,}900 = 14{,}900$

We now need to know the square root of 14,900. In other words, what number multiplied by itself makes 14,900?

If you don't have a calculator, the most straightforward way of working out square roots is guessing, and then improving. If you guess 120 might be the answer, then work out 120 × 120 = 14,400. It's not far off but a bit small, so try 123 × 123 = 15,129. That's a bit bigger than 14,900, so try 122 × 122 = 14,884. You're getting closer, but now try the calculator.

Put in <14,900 √ > and you get 122·065.

So the short cut is just over 122m.

Since Pythagoras proved this theorem over 2,500 years ago, there have been at least 300 *more* proofs involving complex algebra, drawings and trigonometry, and this one, which just involves staring at a few shapes.

To prove $a^2 + b^2 = c^2$

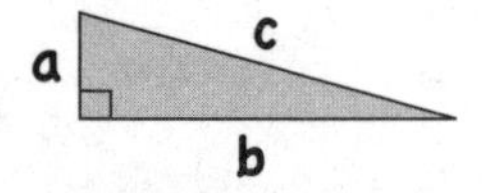

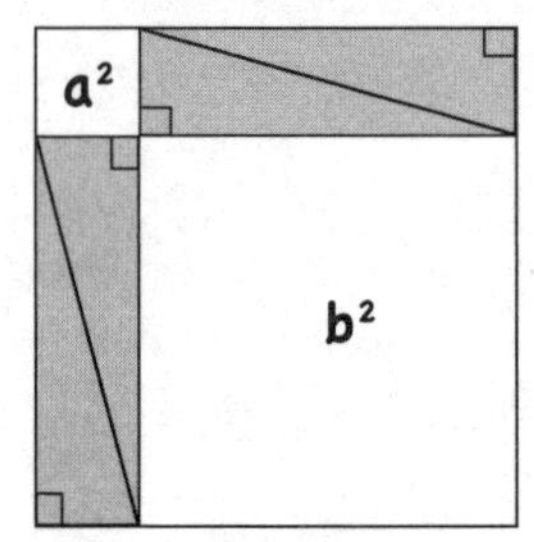

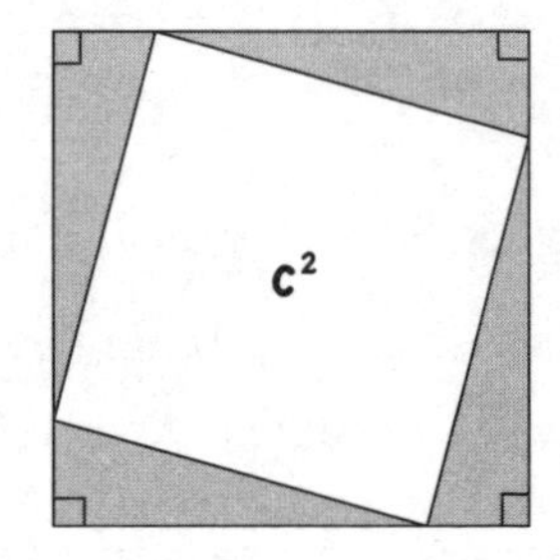

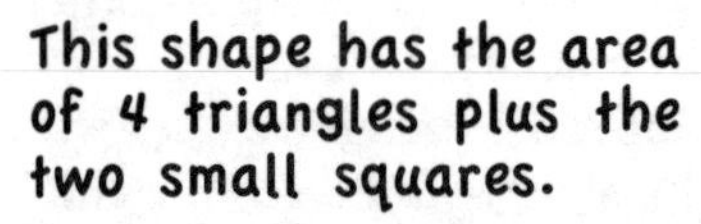

This shape has the area of 4 triangles plus the two small squares.

This shape has the area of 4 triangles plus the big square in the middle.

The outlines of the two lower diagrams are both squares with sides measuring (a+b). This means they have the same area, and therefore the area of the two small squares = the area of the big square. So we have proved that $a^2 + b^2 = c^2$!

HOW DOES CHANCE WORK?

The chance of something happening is measured in a fraction or a percentage. If something is definitely going to happen (e.g. how likely is it to rain at some point next year?), the chance is 1 or 100%. If something is definitely not going to happen (e.g. what's the chance of you growing wings?) the chance is 0 or 0%. If something is equally likely to happen or not happen (e.g. the chance of tossing a coin and getting heads), the chance is $\frac{1}{2}$ or 50%.

When things are very unlikely, it's far more helpful to say something like 'your chance of winning the UK National Lottery jackpot is about 1 chance in 14 million'. If you wanted to be more accurate you could say it's 1 chance in 13,983,816, which is the same as the fraction of $\frac{1}{13,983,816}$. What's really confusing is if you convert it into a percentage: it works out as 0·00000715%.

Dice Chances

Normal dice have six sides so your chance of throwing any particular number is $\frac{1}{6}$ or 16·7%. When you throw two dice, there are 36 different ways they can fall.

Suppose you wanted to throw a total score of 12. You need to throw a double six, and out of the 36 possible ways, there is only way to throw it. Your chances are $\frac{1}{36}$ = about 2·8%.

Suppose you wanted a score of 3. You could either throw 2 and 1 or 1 and 2 (the dice could be either way round), so there are two ways. Your chances are $\frac{2}{36} = \frac{1}{18}$ = about 5·6%.

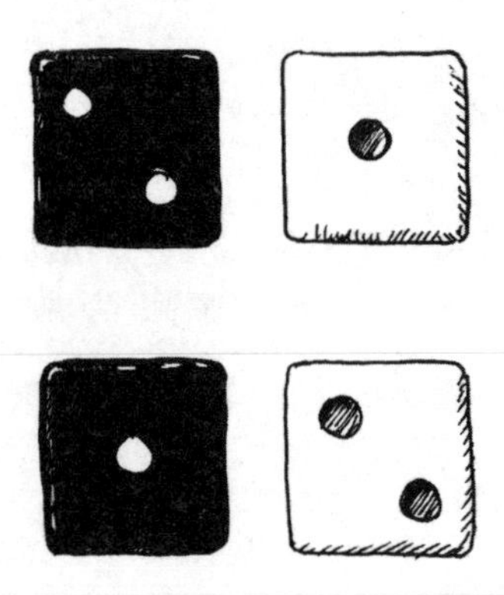

The most common score with two dice is 7 because there are six different combinations that make a score of 7. The chance is $\frac{6}{36} = \frac{1}{6}$ = 16·7%.

Birthday Chances

This is probably the strangest thing in this book: if you have 30 strangers in a room, there is a 70% chance that two of them will share a birthday!

The way to work this out is to start by calculating the chances that *nobody* shares a birthday (just the same day and the month, not the year). You start with Fred standing all alone in the room and then Janet arrives. What is the chance she has a different birthday from Fred? For these sums we'll say there are 365 days in the year and ignore leap years because they hardly affect the answer but they make the numbers awful!

The chance of Janet having the same birthday as Fred is $\frac{1}{365}$. Therefore her chances of *not* having the same birthday are $\frac{364}{365}$.

In comes Barney, and so long as Janet's birthday is different from Fred's, then his chances of *not* sharing a birthday with either of them are $\frac{363}{365}$. The chances that none of the three share are:

$$\frac{364}{365} \times \frac{363}{365} = 99{\cdot}18\%$$

Along comes Agnes, and her chances of *not* sharing a birthday are $\frac{362}{365}$, so the chances of all four of them being different are

$$\frac{364}{365} \times \frac{363}{365} \times \frac{362}{365} = 98{\cdot}37\%$$

Gradually the room fills up and we keep multiplying more and more fractions to find the chances of nobody sharing a birthday. Something strange happens when the twenty-third person comes in. The sum to find out the chances of nobody sharing is:

$$\frac{364}{365} \times \frac{363}{365} \times \frac{362}{365} \times \ldots$$

$$\textit{and so on} \ldots \times \frac{345}{365} \times \frac{344}{365} \times \frac{343}{365} = 49{\cdot}27\%$$

This means the chance of *nobody* sharing is now less than 50%. Therefore the chance that two people do share a birthday is slightly more than 50%, so it's more likely than not!

By the time you've got 30 people in the room, the chance of nobody sharing is down to about 30%, and therefore the chance of two people sharing a birthday is about 70%. If you find it hard to believe then next time you've got 30 people together, ask around. Strange but true.

Card Chances and Poker Hands

If you have a normal pack of 52 playing cards, it's often useful to know how likely you are to get various combinations of cards. Some of the chances are easy to work out for ourselves.

What are the Chances of the Top Two Cards Being a Matching Pair?

If you shuffle the pack and then turn over the top card, it could be anything, e.g. the 4 of clubs. There are three more cards left in the pack that would match it, e.g. the 4 of hearts, 4 of spades or 4 of diamonds. There are 51 cards left in the pack. Therefore the chance of the next card being one of the three you need is $\frac{3}{51}$. This is a fraction you can reduce by dividing 3 into top and bottom, giving you a chance of $\frac{1}{17}$.

In other words if you keep shuffling a pack of cards and then checking the top two cards, on average you should get a pair one time out of every seventeen times you do it.

What are the Chances of Being Dealt Five Cards in the Same Suit?

If you're playing poker, this would be a flush and you would be extremely pleased! But how likely is it?

The first thing to realize is that it makes no difference whether you simply take the top five cards off a shuffled deck, or if you're sitting with a bunch of players and you get your cards dealt one at a time in turn with the others. So we'll imagine you just take the top five cards off a deck.

The top card can be anything. Obviously it's the same suit as itself so the chance of this is 1 (or 100%). Let's say we get the 7 of diamonds. Of the 51 cards remaining there are 12 cards left in the same suit. Therefore the chance of the second card being the same suit is $\frac{12}{51}$.

For the third card there are 11 cards of the same suit out of 50 left, so the chance is $\frac{11}{50}$. Likewise the chance of the fourth card matching is $\frac{10}{49}$ and the fifth card is $\frac{9}{48}$. To get the answer for all five cards matching we multiply these chances together:

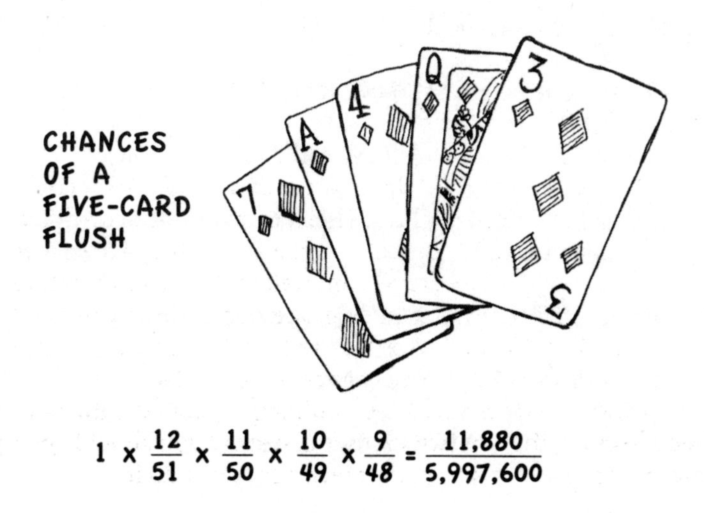

$$1 \times \frac{12}{51} \times \frac{11}{50} \times \frac{10}{49} \times \frac{9}{48} = \frac{11,880}{5,997,600}$$

It's an ugly fraction, so we'll round it off: 11,880 is very close to 12,000 and 5,997,600 is very close to 6,000,000. This leaves us with:

$$\frac{12{,}000}{6{,}000{,}000} = \frac{1}{500}$$

So your chance of being dealt a 5 card flush straight off is about 1 in 500 or 0·2%.

Poker Hands

In poker, the best hands are the ones that are least likely to turn up. Here they are in winning order:

❶ 1 in 650,000: ROYAL FLUSH (A, K, Q, J, 10 all the same suit)

❷ 1 in 72,000: STRAIGHT FLUSH (a run of five cards all in the same suit, e.g. 7, 8, 9, 10, J all hearts)

❸ 1 in 4,000: FOUR OF A KIND

❹ 1 in 700: FULL HOUSE (three of a kind and a pair)

❺ 1 in 500: FLUSH (all five cards the same suit)

❻ 1 in 256: STRAIGHT (e.g. a run such as 2, 3, 4, 5, 6 mixed suits)

❼ 2%: THREE OF A KIND (e.g. three aces)

❽ 5%: TWO PAIRS (e.g. two eights and two threes)

❾ 42%: ONE PAIR (e.g. two queens)

The Ten-Card Poker Trick

You'll notice that any full house beats three of a kind, and three of a kind beats two pairs. Here's a trick I picked up in Dublin from another maths author called Rob Eastaway. We don't take any responsibility for how you might use it.

You need ten cards from a pack – three sets of three and a singleton.

You're going to play with a friend, and you deal five cards to each of you. The sneaky bit is that you can decide who wins and loses!

The secret is gloriously simple. It doesn't matter how you shuffle the cards up, whoever gets the single card will lose! If you're used to handling cards, it's easy to keep the single card at the top or bottom and so ensure the right person gets it. If you're not so sure, put a tiny fold in one corner so you can see who it's dealt to, and that way you'll know who'll win.

If you're smart, you'll let your friend win a few times, and when he starts to get confident, you can raise the stakes and win back.

Some Random Chances

People love speculating on strange chances. Here are a few, but don't take them too seriously!

- The chances of a clover being four-leafed are 1 in 10,000 or 0·01%.

- The chances of a pregnancy producing more than one baby are gradually getting higher. Currently the chances of twins, triplets or even more are about 3%.

- If you're about to throw a dart at a dart board, but then somebody blindfolds you and spins you around, your chances of hitting the board are about 2%. The chances of hitting the bullseye are about 1 in 100,000. But please don't test this one.

- The chances of getting a hole-in-one at golf are supposedly 1 in 5,000.

- Your chances of being struck by lightning are about 1 in 3,000,000. Coincidentally, these are the same chances as meeting an alien.

- The chances of an asteroid hitting Earth sometime in the next century are 1 in 5,000. And if a great big dirty asteroid does hit Earth, what are the chances you've just hung your washing out? About 100%.

- What are the chances of getting full marks on a multiple choice exam purely by guessing? If there are 30 questions, each with four possible answers, the chances are 1 in 4^{30} = 1,152,921,504,606,846,976. If you only need to score 50% or more to pass, your chances of victory are much harder to work out, but it comes to about 1 in 364. The good news is that your chances of guessing every single question wrong are only 1 in 5,600.

Two Misleading Chances

People can get very confused by chance, and sadly the world has lots of nasty, unscrupulous characters who like to take advantage of gullible punters by offering them unfair bets. If you'd like to be one of those nasty, unscrupulous characters then here are two little tricks to delight your soul. The first trick makes the punter think he's clever and the second will clean him out.

The Black and White Cards

You're sitting at a table with poor old Malcolm. You show Malcolm that you have three cards. One is black on both sides, one is white on both sides and the third has one black side and one white side.

Let Malcolm shuffle the cards under the table, draw one out and place it on the table so that neither of you can see the other side. The other two cards stay out of view. The side of the card you can both see is black.

'It's obviously not the white on white card,' you say. 'So it's either the white on black card or the black on black card.' Malcolm nods wisely. This is true.

'So if it's one or the other, it must be an even chance.' Malcolm nods wisely again. 'I'll bet you £1 the other side is black.'

'No thanks,' replies Malcolm. He's suspicious but isn't sure why.

'Oh come on,' you say. 'I'll tell you what – if the other side is black, you give me £1, but if it's white I'll give you £1·50. How's that?'

To Malcolm this sounds too good to resist, so he gets his money out . . . and the chances are that two times out of three you'll win. In other words, if you play this game three times, on average you'll give Malcolm £1·50 and he'll give you £2.

This is the secret: whatever colour is first showing, you *always* bet that the other side is the same colour. Two of the cards have matching colours but only one card is a mix of white and black. Malcolm only has one chance in three of the colour being different.

If Malcolm thinks long enough he might work this out, so it's time to move on to trick two.

The Two-Coin Trick

This trick is so simple, and yet so strange! It's perfect if you happen upon Malcolm sitting with his girlfriend Sandra. Sandra is going to help you take some money off Malcolm; all she has to do is follow your instructions and not whisper clues to Malcolm.

Here are two coins. Shake them up then drop them on the table so that only you can see them.
OK.
If it's two TAILS shake them up again.
It's not two tails.

Can you show us that ONE of the coins is heads? Keep the other coin hidden.
There!
H
So, Malcolm, the other coin can be HEADS or TAILS – yes?
Yes.

I'll give you £1·50 if it's HEADS but you just give me £1 if it's TAILS.
EH? But surely it's an even chance of heads or tails?
H
Mystery coin still hidden
OK! You've got a bet!

So, does Malcolm get a good deal? Of course not. In fact, once again you should win this bet two times out of three. The confusing bit is that when you throw two coins it seems like they can only land in three different ways: two heads, two tails or one of each. However if you use two different sized coins you'll see there are four possible ways they can land:

You asked Sandra to shake the coins up again if it was two tails, so you eliminated that possibility. This means that when you got round to betting, there were only three ways left. When Sandra shows one head, two of the ways have the other coin as tails! That why you'll win two times out of three.

The Bookie's Profit

Suppose you have 12 balls in a bag and only one of them is black. You have to shut your eyes and pull out one ball. If it's the black ball then you win, but what are your chances of winning? It's 1 out of 12, which we can write as a fraction $\frac{1}{12}$.

Another way to describe this is to say there are 11 ways of *not* picking the black ball against one way of winning. This gives the odds *against* winning as 11 to 1 which bookies usually write as 11/1. This is how bookmakers work out their bets.

A bookie who didn't want to make a profit would offer you odds of 11/1 *against* you picking the black ball. If you bet £1 and failed, he would keep the £1. If you bet £1 and won then he would give you back your £1 along with the £11 you have won.

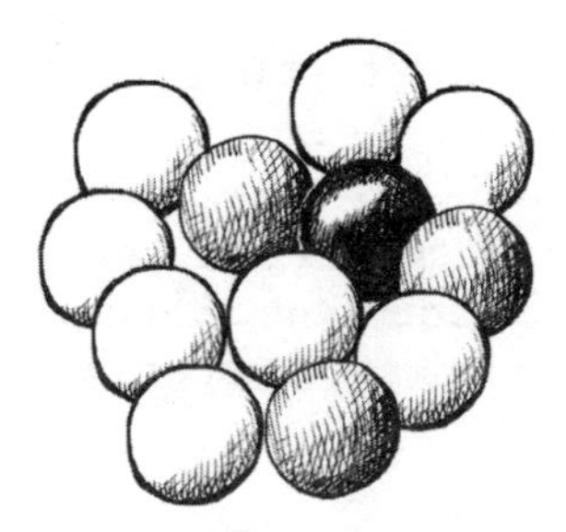

Chances of picking:	Bookie's odds:
BLACK $\frac{1}{12}$	11/1
WHITE $\frac{8}{12}$	1/2

↑ Number of ways you CAN'T win ↑ Number of ways you CAN win

Suppose you worked through the bag, picking out the balls one at a time. You know you'll lose 11 times and win once. If he offered you 11/1 each time, after you'd picked out the last ball, you would have paid him 11 × £1 = £11. He would have paid you 1 × £11 = £11, so it's fair.

You decide that picking the black ball is too unlikely, so instead you decide to try to pick out one of the 8 white balls. Your chance of winning is $\frac{8}{12}$. The bookie says you've got 4 to 8 chances against winning, so his fair odds are 4/8, which reduces to 1/2. If you bet £1 and manage to pull out a white ball, you'll win $\frac{1}{2}$ × £1 = 50p

How Do You Convert the Bookie's Odds into Chance?

Our bookie is also offering 3 to 1 against picking out one of the grey balls. If we want to check that he's being fair, we need to convert his odds into the chance of picking a grey ball and see if it's right.

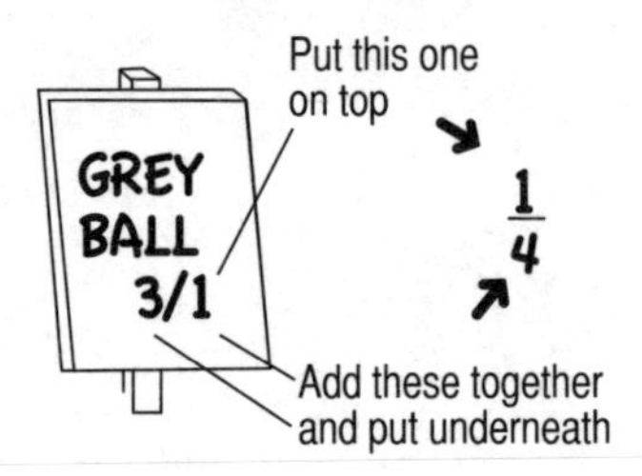

The bookie's odds suggest the chance should be $\frac{1}{4}$. As there are 3 grey balls out of a total of 12, that makes a chance of $\frac{3}{12}$, which is $\frac{1}{4}$. Therefore he is being fair!

Here's the clever bit. Let's suppose we didn't know how many balls were in the bag, we just knew they were white, grey or black. We can see if the bookie is being fair simply by looking at all his odds, converting them into chances and adding the chances for all the different outcomes together.

Black odds = 11/1, so chances = $\frac{1}{12}$

White odds = 1/2, so chances = $\frac{2}{3}$

Grey odds = 3/1, so chances = $\frac{1}{4}$

If he is going to be absolutely fair, then these chances should add up to 1. You can either add the three fractions, or use a calculator and convert to decimals and add, but either way you'll find these chances add to exactly 1. This is a very generous bookie. Unfortunately, he doesn't exist.

Sporting Bets (And the Chances of Elvis Working in a Chip Shop)

When it comes to sporting bets, you can't analyse the chances as easily as picking balls out of a bag. What's more, a bookie is never 'fair' as we've described, because he has to make a profit to pay for his old sports car, his chunky gold bracelet and his villa in Portugal.

Let's check on 'Honest Sid' and see how much profit he hopes to make.

HONEST SID

CUP FINAL	
ALBION	5/4
ROVERS	EVENS
DRAW	11/2

First we convert Sid's bets to chances: 5/4 is a chance of $\frac{4}{9}$ or 0·444, 'evens' is the same as 1/1, which is a chance of $\frac{1}{2}$ or 0·5, and 11/2 is a chance of $\frac{2}{13}$ or 0·154. If we add up all the decimal chances we get a total of 1·098.

This answer tells us that for every £100 Sid pays out, he expects to take £100 × 1·098 = £109·80. That's a profit of £9·80.

Some bookies also offer unreal bets, such as Elvis being discovered alive and well and working in a chip shop. You're better off buying a lottery ticket; although the chances of winning the jackpot are only 1/13,983,816, in reality it's more likely to happen. As the King himself would have said 'Well, it's a-one (in about fourteen million) for the money . . .'

EXTRA MATHS

You've made it this far: congratulations! How about showing off by proving you can handle some more theoretical stuff? The following two sections involve maths that you might not use everyday – or even at all – but it's interesting to know the basics about them, especially if they gave you nightmares at school!

Angles, Triangles and Trig

The angle where two lines meet is measured in degrees, which are followed by a little sign, °. If you want to see how big an angle of 1° is, get a long piece of string, loop it round your thumb and stretch your arm to the side. With your other hand, pinch the loose ends together in front of your face. The angle where the two ends of string meet will be about 1°.

The corner of a square is 90º and known as a *right angle.* If you spin round one complete turn, you turn through 360º. An angle of 180º is a straight line and the three angles in a triangle always add up to 180º. If you cut any triangle shape out of paper and tear off the angles, you can put them together and they will make a straight line.

The four angles in any four-sided shape always add to 360º, so if you put them together they will fit perfectly.

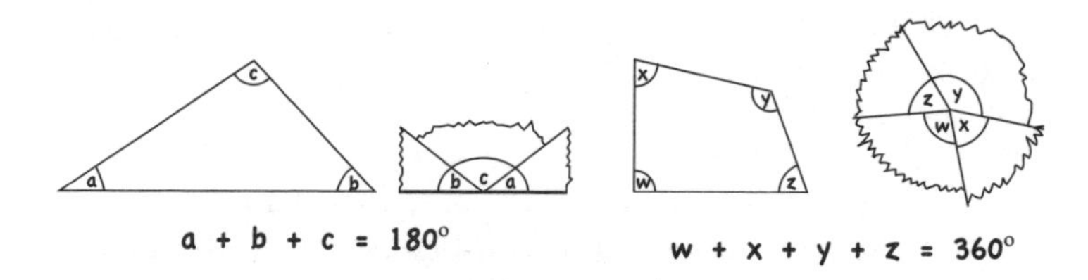

Have you got a calculator with a pile of mysterious buttons that you don't use? It seems a shame when you've paid for them so here's a very quick introduction to sin, cos and tan.

The basic idea is that if you have a triangle and know the lengths of one or two sides and angles, you can use *trigonometry* to work out the others. The easiest triangles to deal with are right-angled because, apart from the right angle, you only need to know one more side length and one more angle and you can work the rest of the measurements out.

If you know what one angle is, then if you get the side *opposite* the angle and divide it by the long side, which is called the hypotenuse, you get a fraction. This fraction is called the *sin* of the angle. (Don't get too excited: *sin* is pronounced *sine* and it doesn't have much to do with do naughtiness.)

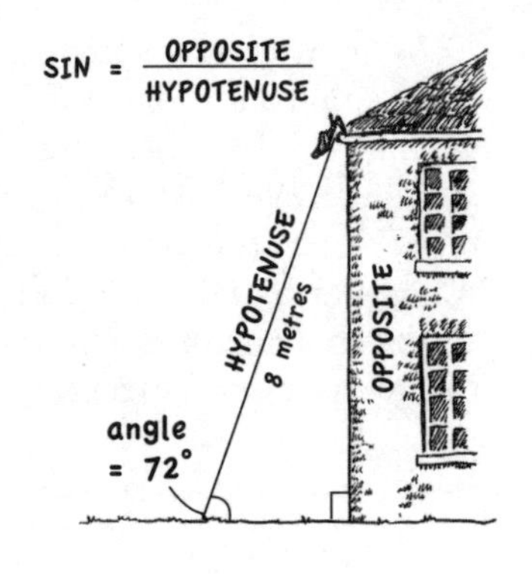

Let's say that you're trying to retrieve one of your best shoes out of a high gutter. (Goodness knows why these things happen but we all know they do.) You have an 8-metre ladder leaning against the building.

The ladder, the wall and the ground make a right-angled triangle. If you measure the angle at the ground and find it's 72º, you can go on to work out the height of the gutter so you know what to tell the ambulance man later.

The 8-metre ladder is the hypotenuse of the triangle so *hypotenuse* = 8. The height we want to work out is the *opposite* side to the angle of 72º, so we put these bits in a little equation:

$$\sin 72^\circ = \frac{\text{opposite side}}{8}$$

Multiply both sides of the equation by 8 and we get

$$\sin 72^\circ \times 8 = \text{opposite side}$$

To work this out you grab the calculator and put in <sin 72 => which gives you 0·951

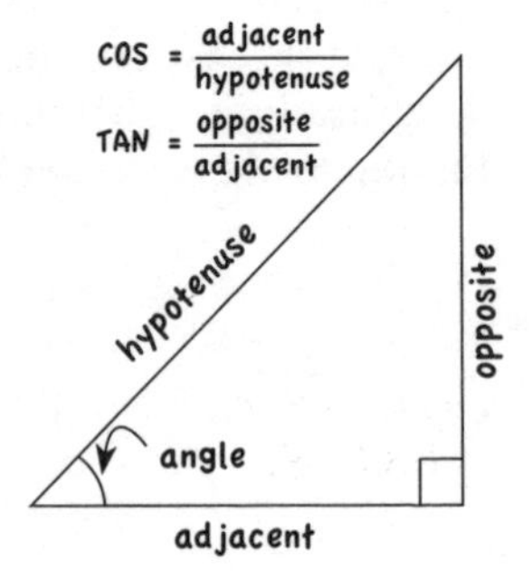

You then multiply this by 8 to get the answer 7·608. That's how many metres high the gutter is!

Cos and tan are the other fractions you can make if you combine other sides of the triangle.

And that's pretty much all that anybody ever needs to know about trig . . .

What the Hell's a Logarithm?

Whenever people are talking about the darkest and most depressing bits of maths, logarithms usually get a mention. For many people it conjures up nightmares of meaningless numbers and sarcastic teachers. But now that it doesn't matter, would you like to know what it was all about? There are no tests, no reports and no flying board rubbers. The beast can't hurt you.

Logarithms were invented in 1614 by the Scotsman John Napier and for 350 years (until calculators were invented) they were the only reliable short cut to multiplying or dividing huge numbers. But how do they work?

Here's a fairly straightforward sum:

1,000 x 100 = 100,000

We could also write this sum as $10^3 \times 10^2 = 10^5$, which is exactly the same thing, but instead of multiplying the big numbers, we just added the powers: 3 + 2 = 5. John Napier realized that you could convert *any* number to powers of ten, and then you could multiply or divide the numbers just by adding or subtracting their powers.

This is where it gets fiddly because most of the powers involved are not nice whole numbers, for instance $78 = 10^{1{\cdot}89209}$. When the powers become fancy decimals, they get known as *logarithms*. As $78 = 10^{1{\cdot}89209}$ we can say that the logarithm of 78 is 1·89209.

Converting numbers to logarithms is an extremely tedious process, but a friend of Napier's called Henry Briggs worked out a set of conversion charts called log tables. Some log tables only gave three places of decimals so 78 would have come out as $10^{1{\cdot}892}$. Henry Briggs' best set of tables would have given you $78 = 10^{1{\cdot}89209460269048}$. Obviously, more accurate logarithms give more accurate answers. (Isaac Newton worked to 50 decimal places when he was studying how the planets and stars moved but he was obsessive.)

Now let's do a sum with them.

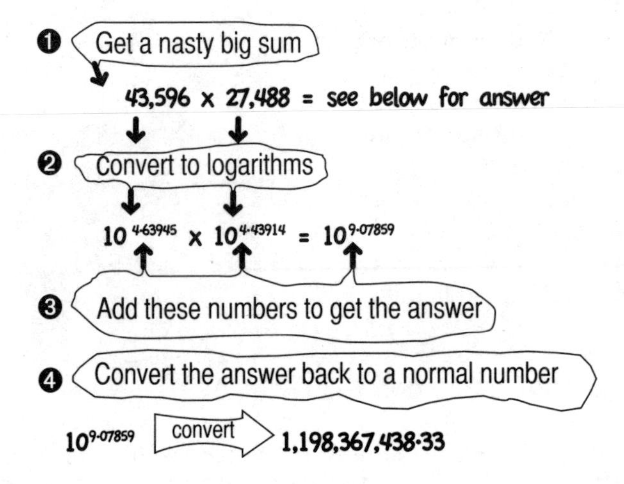

The real answer is = 1,198,366,848. The error is about 1 in a million!

A Magic Short Cut to Roots

You can find the square roots or cube roots of numbers simply by dividing the logarithm by 2 or 3.

If you were Isaac Newton and you wanted the cube root of 591, first you would check the log tables and find $591 = 10^{2{\cdot}771587}$. Next you would work out $2{\cdot}771587 \div 3 = 0{\cdot}923862$. Finally, you'd convert $10^{0{\cdot}923862}$ back to get the answer 8·391942. (If you multiply $8{\cdot}391942 \times 8{\cdot}391942 \times 8{\cdot}391942$ you'll find it makes 591.)

Not only is that the right answer, but logarithms have also saved you hours of mind-numbing arithmetic!

GLOSSARY

There are lots of words to describe different things in maths, but there's enough to take on board in this book without having to learn a new language, so I've tried to avoid them as much as possible. However, here is a handy guide to the main terms:

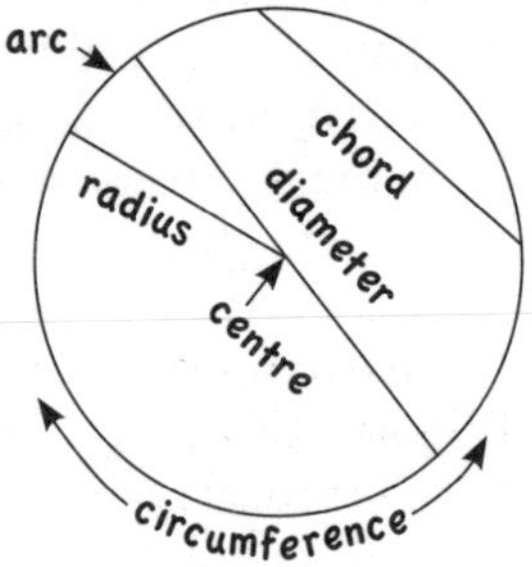

Acute – an angle that is less than 90° (or sharper than a right angle).

Arc – a section of the edge of a circle. It can be anything from a tiny bit to almost a complete circle.

Chord – a straight line across a circle that doesn't go through the centre.

Circumference – this can be the outside edge of any area, but it usually refers to the distance around the edge of a circle. It's also the punchline to a really poor joke: 'Which of King Arthur's knights was good at maths?'

Co-efficient – a number by which you multiply another number (or even a bracketful of numbers). If you have $3(2x + 7)$ then 3 is the co-efficient of the bracket and 2 is the co-efficient of x.

Composite – any number that isn't prime. In other words it will divide by numbers apart from 1 and itself.

Crooked Waiter answer (from page 36) – you must forget the £30! In the end the ladies paid £27, of which £25 went in the till and the other £2 went to the waiter.

Decimals – short for 'decimal fractions', in other words a string of digits with a decimal point somewhere in the middle such as 0·667 or 365·26.

Degree – a measurement of angles that uses this little sign °. Temperature is also measured with degrees (of either Kelvin, Celcius or Fahrenheit).

Denominator – the bottom number in a fraction. If you have $\frac{4}{7}$ then 7 is the denominator.

Diameter – the line across the middle of a circle.

Dividend – in a dividing sum, it's the number you're splitting up. If we have $35 \div 5 = 7$, then 35 is the dividend. If a company is doing well, every so often they pay out a dividend, which is a load of money that gets divided up between the shareholders.

Divisor – In a dividing sum, it's the number you're dividing by. In the sum $48 \div 4 = 12$, the divisor is 4.

e – this is a special number: 2·71828183. It comes up when people are calculating growth. It applies to living things like plants, but also when banks are calculating the interest on money.

E – on a calculator this means 'exponential' which means you have to multiply the figure displayed by a power of 10.

Ellipse – this is like a circle with two centres. These 'centres' are called focal points and the further apart these points are, the more squashed the ellipse looks. Planets go round the Sun in elliptical orbits, but the Earth's orbit is very close to being a circle.

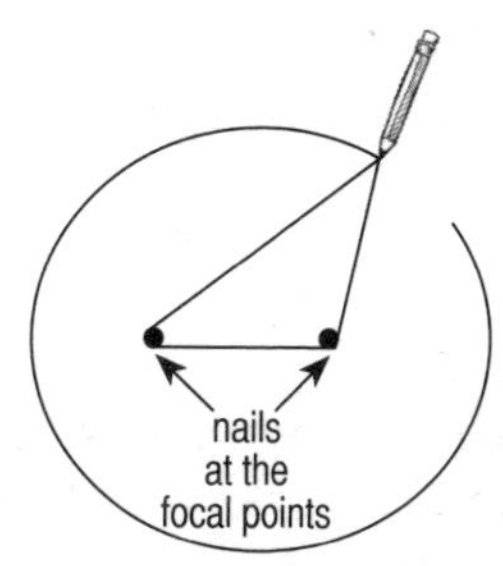

HOW TO DRAW AN ELLIPSE

Bang in two nails. Put a loose loop of string around them, then draw round with a pencil, keeping the string tight.

Equilateral – an equilateral triangle has all three sides the same length. This makes the angles all exactly 60°.

Expand – this means to get rid of brackets, usually in algebra. If you have $4y(3 - 2y)$ it expands to make $12y - 8y^2$.

Factorial – this is when a number is multiplied by all the smaller numbers down to 1. It has a special sign: '!'. For example $4! = 4 \times 3 \times 2 \times 1 = 24$. If you have four horses racing, this tells you the number of different orders they can finish in. Unfortunately it won't tell you which one will come first.

Factorize – this happens in algebra when you put things into brackets. If you have $6x^2 + 9x$, both bits will divide by 3x, so you can factorize it into: $3x(2x+3)$.

Factors – these are all the whole numbers that divide into another number. The factors of 60 are 1, 2, 3, 4, 5, 6, 10, 12, 15, 20 and 30. There are also prime factors. This is the set of prime numbers that you need to multiply together to make a composite number. The prime factors of 60 are $2 \times 2 \times 3 \times 5 = 60$.

Irrational – a decimal fraction where the digits go on for ever and never repeat in a predictable pattern.

Isosceles – an isosceles triangle has two sides the same length, and two of the angles will be the same.

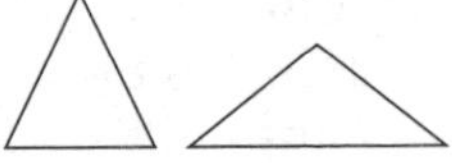

LHS – abbreviation that stands for the left-hand side of an equation.

Mean – the sort of average where you add everything up and then divide by the number of things you've added.

Median – if you have a range of results, the median is the one in the middle.

Mode – if you have a range of results, the most common result is the mode.

Numerator – the top number in a fraction.

Oblong – a shape that is longer than it is wide, so it's usually a rectangle although it can be an ellipse.

Obtuse – an angle bigger than 90° but less than 180°. An obtuse triangle is a triangle that has an obtuse angle in it.

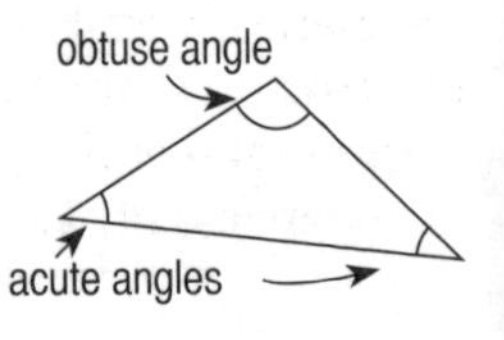

Perpendicular – a line that is at a right angle to another line or surface.

Pi (or π) – 3·14159265 . . . The special name for the number you get if you divide the circumference of a circle by the diameter.

Power – this is when a number is multiplied by itself. 4^5 is four to the 'power' of five = 4 × 4 × 4 × 4 × 4 = 1,024.

Prime – a number that only divides by itself and 1.

Product – what you get when two or more numbers are multiplied together. The product of 4, 7 and 8 is 4 × 7 × 8 = 224.

Protractor – a device to measure angles, usually about the size of a CD snapped in half with lots of numbers round the edge. Also quite handy for scraping ice off car windscreens.

Quadratic – this usually refers to a quadratic equation. This is a bit of algebra with a squared term in it such as x^2. Quadratic equations usually have two different answers.

Quadrilateral – any shape with four straight sides.

Quotient – the answer to a division sum such as 14 ÷ 2 = 7. Here 7 is the quotient.

Radius – the distance from the centre of a circle to its edge. The radius is always half as long as the diameter.

Rational – a decimal fraction where the same pattern of digits keeps repeating.

Reduce – to make the numbers in fractions smaller.

Reflex – an angle bigger than 180º.

RHS – abbreviation that stands for the right-hand side of an equation.

Right angle– an angle of 90º, which is the same as a square corner. It's usually marked with a little box sign.

right angle

Rounding off – making a complicated number look simpler.

Scalene – a scalene triangle has all three sides different lengths.

Sector – a shape like a slice of pizza.

Segment – either part of a straight line, or a bit chopped from a circle with a straight line.

Set square – plastic triangle thing found in geometry sets. There are usually two sorts. The fatter looking version has one right angle and two angles of 45°. The skinnier one has a right angle, and angles of 30° and 60°. Apparently 90% of set squares are given to kids when they start secondary school by well-meaning aunts.

Simplify – this usually refers to starting with some complicated algebra and making it simpler. If we had this: $3(2x - 4) + 5(1 - x)$ we can simplify it by multiplying it out to get this: $6x - 12 + 5 - 5x$ which then becomes $x - 7$.

Square/square root – if you 'square' a number you multiply it by itself, e.g. $7 \times 7 = 49$. Working backwards, the 'square root' of 49 is 7.

Sum – the 'sum' of a set of numbers is the same as the total; in other words, what you get if you add them up.

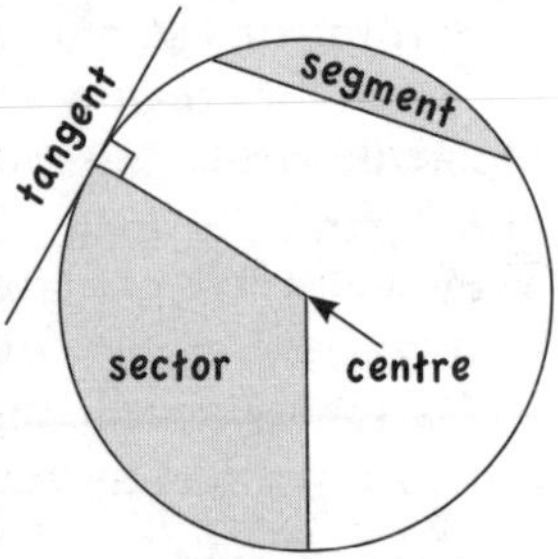

Tangent – a line that touches a circle at one point. If you draw a radius of the circle to touch the tangent, they always meet at 90°.

Vulgar fractions – this is the proper term for fractions with one number over another such as $\frac{2}{3}$, as opposed to decimal fractions, e.g. 0·618.

Zero – this is probably the most complicated number of all, because people aren't even sure if it is a number. If it is a number it causes all sorts of problems, especially when you try to divide with it. If it isn't a number, then how come it can be made from numbers, e.g. $2 - 2 = 0$?

QED

These letters turn up at the end of maths proofs and stand for the Latin phrase *quad erat demonstrandum,* which means 'that which was to be demonstrated'. It's a convenient way of saying 'There, see? I've proved it works, OK?'

Hopefully by now I've QEDed how maths works for you and shown how all the different bits link up. All that's left is to wish you the best of luck the next time you're faced with a pile of numbers! As my mate Blakey said, 'I've spent most of my life not understanding it, and now I can't see what my problem was.'

Here's one last cute trick:

❶ Pick any four-digit number with different digits. A nice way of doing this is to get nine playing cards ranked ace (as 1) to 9, shuffle them up and then pick out four at random. Let's say you've got the number 4,728.

❷ Turn the digits round: 8,274. Subtract the smaller number from the bigger one: 8,274 – 4,728 = 3,546.

❸ The digits of the answer will always add to 18: 3 + 5 + 4 + 6 = 18.

Did you want to know why? I daresay it can be proved with algebra, but there's a much nicer reason why it works. It's magic.

ACKNOWLEDGEMENTS

My thanks to the fantastic team at Michael O'Mara who made this book possible, as well as Andrew Pinder for his artwork, Richard Collins for his proofreading, and especially to my editor, Kerry Chapple, without whom this book would never have been so much fun for you to read or for me to write.